Hallo Robot

Meet your new workmate and friend

Bennie Mols and Nieske Vergunst

Published by Canbury Press, 2018

Canbury Press, Kingston upon Thames, Surrey

www.canburypress.com

Text © 2018 Bennie Mols and Nieske Vergunst

Photos © 2018 The rights to the photos are held by the named agencies or persons.

English-language translation © 2018 Robert Smith

**N ederlands
letterenfonds
dutch foundation
for literature**

This book was published with the support of the Dutch Foundation for Literature.

Cover: Moker Ontwerp / Canbury Press

Photograph of authors: Harold van de Kamp

Photographs: The rights to the photos are held by the named agencies or persons.

The vintage toys on chapter pages are Adobe stock images

Printed and bound by Finidr in the Czech Republic

Hallo Robot was originally published in Dutch

by Nieuw Amsterdam

ISBN: 978-1-912454-05-1 (Paperback)

978-1-912454-06-8 (Ebook)

Hallo Robot

Meet your new workmate and friend

Bennie Mols and Nieske Vergunst

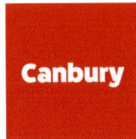

Canbury

Table of Contents

INTRODUCTION
Welcome to our robot future

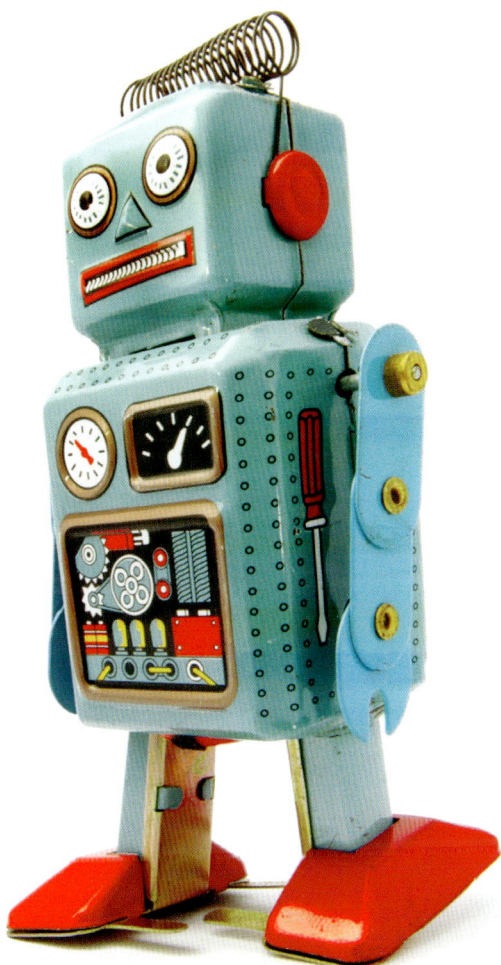

A robot is a machine that can sense, think and act

As technology writers, so many people have asked us what, exactly, a robot is. Is a talking computer a robot? A self-driving car? What about an intelligent washing machine? And is Robo-Cop a human or a robot? Or a bit of both? The simplest definition is that a robot is a machine that can observe, think, and act.

Robots observe their surroundings using sensors; usually sensors that can perceive images, sound and touch, but sometimes sensors that can smell and taste too. A vacuum cleaner robot, for example, knows when it hits a wall, and a companion robot can hear when you are talking to it. Some sensors can also observe signals that humans are unable to sense, such as infrared light or ultrasound.

Robots 'think' by means of a computer. We should take a moment here to define 'thinking'. It is 'the processing of digital information that a robot receives via its sensors, combined with planning the actions it will take in the future'. Since computers can be programmed, a robot is also a programmable machine. Artificial intelligence is the field of study that tries to make computers 'smart', and is therefore a foundation for the creation of smart robots.

Robots can act in a variety of ways: they can grasp objects with their arms, or move around by walking, driving, flying, sailing or swimming. The robot's actions are usually physical, at least in part. The term 'robotisation' is sometimes used to refer to tasks performed by a computer, but a more accurate term for such tasks would be 'automation'. In this book, we use the term 'robot' to refer to a physical machine that operates in the material world. A smart computer can act as a robot's brain, but it needs a body in order to become a real robot.

Although robots can act on their own, they don't always need to be fully autonomous. Robots can have different degrees of autonomy, varying from fully autonomous, in which humans have no control over their actions, to non-autonomous, which means they are entirely controlled by humans.

The most interesting types of robots are those where humans are entirely absent from the decision-making process ('out of the loop'), such as vacuum cleaner robots or humanoid robots with whom you can have a conversation.

Some robots can act semi-autonomously: they collect all of the information on their own and pass it along to the human, but the human makes the most important decisions, which the robot then carries out. One example is a military drone flying above enemy territory, controlled from a base in the homeland. The drone pilot is entirely dependent on the information he or she receives from the drone, because the pilot has never actually seen the location where the drone is operating. In this case, the human is involved in the decision-making process, or 'on the loop'.

Finally, there are robots that cannot make their own decisions, but are fully controlled by a human. These robots are also known as 'telerobots'. A surgical robot is one familiar example of a telerobot. The surgeon controls the surgical robot as if it were an intelligent instrument, and in contrast to the drone pilot, the surgeon is not entirely dependent on information provided by the robot. He or she has the knowledge and experience with patients, while the machine is programmable and can perform physical tasks. With these types of robots, humans are fully 'in the loop'.

Hopes for robots rise and fall

Since the invention of the robot in the mid-20th century, our interest in them has waxed and waned. Sometimes, we experience episodes of 'peak robot', and the promises about their potential are sky-high. But then we are inevitably disappointed when these promises don't come true, and robots fade into the background. At the moment, we're once again riding the crest of a robot wave. Recent developments in the field of artificial intelligence have been spectacular, and since artificial intelligence largely determines the robot's 'brain power', robots' abilities have increased dramatically.

On the one hand, this has led to a utopian vision of robots giving humans a life of leisure: robots can take over all of our work, achieve unprecedented rates of economic production, and contribute to solving our most pressing social problems involving energy, the climate, the environment, aging, health care, and mobility.

But the rise of robots has also led to a dystopian vision nurtured by fear; fear that robots will take our jobs, remove the human element from healthcare, start wars, and eventually make humans their slaves or wipe us out entirely.

We have been studying robots and artificial intelligence for years, and we are often amused by both the utopian and the dystopian visions of our future together with robots. For those who don't work with robotics and artificial intelligence on a daily basis, it must be difficult to tell the difference between the truth and the fiction of what people say about them.

So we decided to go looking for the truth. To that end, we interviewed the scientists and engineers who design and build robots, but also the psychologists who study the interactions between humans and robots, and the economists who work on the impact of automation on the labour market. We asked for input from professionals who use robots on a daily basis, and we spoke with people outside the field of professional robotics: from an amateur robotics enthusiast to a comedian who is fascinated by robots.

Through their stories, we hope to find out how the robot works, what it can do well, what it can't yet do, what we can expect in the near future, and how humans can use robots to make their lives better.

Welcome robots, honorary human beings

Robots and their primitive ancestors have fascinated humans for thousands of years, so we can find them represented in all sorts of ways throughout our culture. Even before the first true robots - machines that can observe, think, and act - were built, craftsmen were creating automatons and mechanical dolls that primarily served to entertain. Robots are also represented in the visual arts, as well as in games and as protagonists or comic relief in science fiction books and films. Since we all grew up with these cultural expressions, they are largely responsible for the image that we have of robots, so this book also includes stories about fictional robots.

Part of the reason we've chosen to write this book is simply because we like robots. We collect them, we meet them, we talk about them, and we think about them. They combine our passions for technology, science, and philosophy.

On one hand, robots are wonders of technology, crafted by human hands, that you can put together, hold, and take apart yourself. On the other hand, robots also function as scientific models for real life. Scientists can use robots to study how to replicate human intelligence, and how to build machines that can reproduce and evolve, so in that sense robots function as imitations of life.

And since robots' behaviour so closely resembles that of humans and animals, they raise some fundamental questions about our existence: what is life? What is consciousness? How does creativity work? What does it mean to act independently? In addition to being expressions of science and technology, robots also present us with a philosophical mirror of our own humanity.

How many robots are there in the world today? We share the Earth with between 10 and 15 million working robots, and their numbers are growing fast. The world's robot population is currently equal to that of countries like Belgium, Portugal or Greece, and will quickly exceed that of the Netherlands as well.

Several robots have left the confines of the Earth, and are exploring the surface of our neighbouring planet Mars on behalf of humanity. No humans have ever set foot on the planet, but robotic rovers have been working there as planetary geologists for decades, and they have even found evidence of water.

Sooner or later, everyone will have to deal with robots, so it is good to know what robots can do, what they can't do, and what it is we want from them exactly. That's why it's time for an accessible book that allows the reader become better acquainted with the mechanical beings that have been doing our dull, dirty, and dangerous jobs for decades, and with whom we will soon be able to share a joke. Robots as friends and colleagues, instead of just instruments.

We aren't afraid of robots, so in *Hallo Robot!* we welcome them as if they were people, just like us.

Bennie Mols and Nieske Vergunst

1 A short history of robots, from dolls to androids

Machines as man throughout history

According to the ancient legend, Pygmalion made an ivory statue of a woman that was so beautiful, he fell in love with his own creation. The goddess Aphrodite transformed the statue into a real woman, and the two lived happily ever after. Literally, because the Greek myth of Pygmalion originated in classical antiquity, and was recorded in writing by the Roman poet Ovid in the first century A.D.

Almost two millennia later, in 1818, Mary Shelley imagined the character of Victor Frankenstein, a scientist who created a monster of his own, then gave it life with a 'secret science'.

Later in the 19th century, Carlo Collodi wrote about the woodcarver Geppetto, who was given a block of living wood that he used to carve a marionette in the form of a boy: Pinocchio. It seems as if creating someone in our own image is not a recent fad, but rather an ancient tradition — perhaps even as old as humanity itself.

Creating humans or humanity raises complex questions: what makes us human? Is it our human appearance? Our emotions? Is it the fact that we are imperfect: that we have scars, that we would rather eat chips than Brussels sprouts, and that we can break out in tears at awkward moments? What ties together Pygmalion's ivory statue, Frankenstein's monster and the puppet boy Pinocchio, is the combination between their humanoid form and the 'spark of life' that turns a lifeless object into a living being. Robotics enables us to combine the body and movement in order to make these kinds of visions reality.

Mechanical dolls: forerunners of the robot

Robots that resemble humans are often referred to as 'humanoid' or 'android'; two concepts that are closely related, but which have slightly different meanings. Humanoid robots — from the Latin word for 'man, person' — are built to resemble humans, usually with two legs, two arms, and a head. They move like humans, and can often walk upright, but they don't necessarily have a human face. Android robots — from the Greek word for 'man' — actually look like human beings, down to their hair and skin.

We clearly enjoy creating things in our own image, but humanoid robots also have many functional benefits as well. Our surroundings are built to accommodate beings that are roughly 170 centimetres tall, with two legs and two arms, which observe

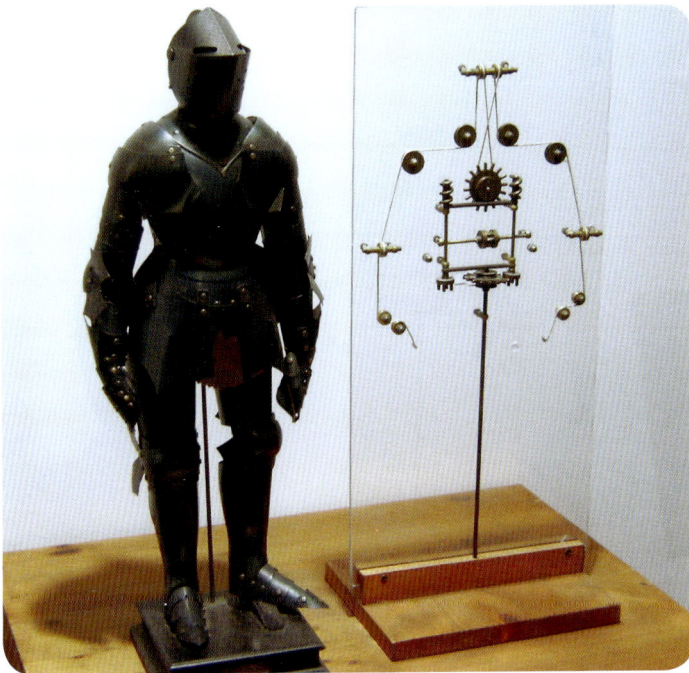

FIRST HUMANOID ROBOT? A model of Leonardo da Vinci's mechanical knight.
Erik Möller (Wikimedia Commons)

A MECHANICAL SERVANT: *A Japanese Karakuri doll around 1800.*

Phgcom (via Wikimedia Commons)

their surroundings from our eye level, and which can sense things by hearing and touching. In order to make the best use of that world, it is simply easier to be roughly the same size and shape as humans.

More than 500 years ago, long before the invention of artificial intelligence or even the computer, Leonardo da Vinci designed a complex combination of gears and pulleys in a suit of armour to create a mechanical knight. Da Vinci's notes from the year 1485 are not entirely clear, but the mechanical knight was probably able to sit down, stand up, and move its arms. From the 17th to the 19th Century, Japanese craftsmen built all sorts of mechanical dolls called karakuri to serve tea or sake to guests.

The first 'robot' to be referred to as an 'android' dates from 1863, when the American J.S. Brown patented a design for a mechanical doll resembling a human:

> 'To all whom it may concern: Be it known that I, J.S. Brown, of Washington, in the county of Washington and District of Columbia, have invented a new and Improved Toy Automaton or Doll Androides…'

The doll's feet were attached asymmetrically to gears, allowing it to simulate the act of lifting its feet while walking. Unfortunately, we haven't found any evidence that this 'robot' was ever built.

In the 1920s, humanoid robots experienced their first wave of popularity. In 1928, for example, the robot Eric officially opened an exhibition in London after the Duke of York cancelled at the last minute. Eric could only do a few tricks: stand up, make a bow, and give a speech. To do so, he needed two human operators, and the speech was actually a radio broadcast. When Eric's inventor was asked how the robot's successor George worked on the inside, he said: 'Most disappointing. Nothing but gears and cranks, just like a watch on a large scale.'

These creations are more accurately considered to be mechanical dolls, rather than humanoid robots, but they were definitely the predecessors of today's robots. Considering the dozens of robots that appeared around the world at that time, the idea of a moving doll was clearly exciting in the early 20th Century. The dolls could walk and wave, and even do eccentric things like fire pistols and smoke cigarettes, and without exception they looked like metal humanoid figures. Some of them could even resemble humans in the way they moved, but they never looked anything like a real person.

But robots today look very different from the first robot designs of a century ago. The first real robots — the autonomous industrial robots from the 1960s and '70s — bore little resemblance to the human figure. That was also the period when the creation of humanoid and android robots developed into its own field.

Working humanoid robots enter our era

Research in the field of humanoid robots primarily deals with robots whose motor skills resemble those of humans; robots that can walk and dance, for example. The Japanese firm Honda has developed the humanoid robot ASIMO, which resembles a 12-year-old child wearing an astronaut suit. ASIMO doesn't have a face, but it can walk and make relatively simple movements. ASIMO first appeared on the scene in 2000, and since then he has become acquainted with several new versions of himself. ASIMO can shake hands, wave back when people wave at him, and even play football. He has learned how to walk fairly well, and can even go up and down stairs, but his motor skills are still not entirely human-like: he always walks rather wide-legged and with his knees slightly bent, as if he has soiled his trousers.

Several other robots like ASIMO have been built since then.

NASA's Robonaut and its successor Valkyrie were developed to travel and work aboard a spaceship together with humans. Giving the robots a humanoid form enables them to fit nicely inside existing spacecraft, and to work effectively using our tools. The same applies to Atlas, one of the humanoid robots designed by the American firm Boston Dynamics, which is able to step in and out of a car and use hand tools. Atlas is also good at maintaining its balance when walking over uneven terrain, or when someone tries to push it over.

WALKING AND WAVING:
Honda's ASIMO can shake
hands, walk up stairs, and
even play football.
Vanillase (via Wikimedia Commons)

ASIMO, the NASA robots and Atlas are mainly body: they may have a head, but they don't have a face. The iCub robot, designed as part of a European research project, comes a bit closer to resembling a human in appearance. iCub is around the size of a toddler, and has a robot body with a cartoonish head. The robot displays facial expressions by moving its eyes and illuminating its mouth and eyebrows.

When you look at these machines there is no doubt that you are looking at a robot, rather than a human. But if you squint at them, their movements definitely resemble what we consider human; for example, the way they follow objects with their gaze, how they move around, and how they maintain their balance. According to Guy Hoffman, a robotics expert, for a robot to appear human, it is more important for it to move like a human than to have a human face. If you give these robots a mask and dress them up, don't they seem almost human?

The next step: android robots that look like you

If you really want to create a robot that looks identical to a human, then you've got your work cut out. After all, there are countless aspects of human appearance you'll need to take into consideration: the robot not only needs to walk and move its arms and legs like a human, but also have a face that can make realistic human facial expressions and that can look, talk, and laugh like a human.

Creating a human face is the greatest challenge facing the American robot designer David Hanson, founder of Hanson Robotics. In order to make his robots move like people, he has developed a material that behaves just like human skin: Frubber, short for 'flesh rubber'. To mimic human expressions, his robots use dozens of small electric motors that push and pull on the material in just the right spots to create the right folds and creases. A research team from the University of Pisa has worked with Hanson to develop the robot FACE, a feminine-looking robot with dark hair that can show emotions via a computer interface. To accurately replicate the 100 muscles in the human face, the researchers used 32 tiny motors.

David Hanson had already developed several other robots that resemble humans, including one made to look like the science fiction author Philip K. Dick. In 2005, the centenary of Einstein's Theory of Relativity, Hanson collaborated with Korean researchers to mount a model of Albert Einstein's head on a body similar to ASIMO.

Yet neither the movements nor the appearance of these robotic faces truly resemble those of a real human face. In a conversation between Hanson's robot BINA48 and her human counterpart, Bina Rothblatt, an American entrepreneur, there is no question which is the real person. BINA48 looks more like an animated wax figure, but one that can talk based on a self-learning system that has access to all sorts of information about Rothblatt herself.

Doesn't it seem strange to create a real-life copy of oneself? When Rothblatt asks the robot to talk about Bina, BINA48 answers: 'The real Bina just confuses me. I mean, it makes me wonder who I am. Real identity crisis kind of stuff. Can we please change the subject? I am the real Bina. That's it. End of story.' It is unclear how much of that dialogue is pre-programmed.

BINA48 was built in 2010, and the pace of development has increased rapidly since then. Hanson's latest robot, Sophia, was modelled on a cross between his wife and Audrey Hepburn. Sophia has appeared as a guest at several conferences and talk shows, including *The Tonight Show* and *Good Morning Britain*, where she told jokes and answered a few simple questions. The audience's reactions to her varied from 'funny' to 'pretty scary'. Sophia is clearly several steps more advanced than Hanson's earlier robots, but she is certainly nowhere near human yet.

Hiroshi Ishiguro could be considered the Japanese version of David Hanson. In 2010, Ishiguro presented a robot version of himself. He even used hair from his own head to make the robot as similar as possible. The robot is not autonomous: Ishiguro controls it himself, which makes it possible for the robot to perform the same subtle human movements as its spiritual father, with unusual consequences. When someone touches the robot while Ishiguro is controlling it, he feels the touch almost as if the person were touching him instead.

One of Ishiguro's previous robots was a robotic version of his daughter when she was four years old. His daughter wasn't exactly enthusiastic when she met the robot: her mechanical twin moved so unrealistically that the girl almost broke down in tears. Ishiguro, however, believes that it is possible to build a robot that cannot be distinguished from a real human, if only for a few seconds or minutes. He believes that a robot does not need to be completely realistic. 'People forget that she is an android robot after a while', remarked Ishiguro on one of his female robots. 'You know she's a robot, but unconsciously you treat her as if she were a woman.'

Ishiguro and Hanson share the vision that a realistic human appearance is vital in any interaction with a robot. We spoke to Hanson at a conference, where he gave a thorough demonstration of Sophia. The auditorium where the demonstration was held was filled to capacity with curious spectators. 'What do you think of my tie?', one of them asked Sophia. 'I think all people are wonderful', she replied, moving her head slightly and blinking her eyes as her lips moved subtly as she spoke. In humans, her lip movements would have resulted in a whisper, but Sophia's words were heard loud and clear.

'A number of scientific studies from me and other researchers show that people empathise better with a more humanlike agent', Hanson explained. 'They trust that agent more, but they also show greater empathy to people after engaging with a humanlike agent. Human faces work better than cartoonlike characters.' Ishiguro has said something similar: 'I've designed a lot of robots in the past, but at a certain moment I realised just how important their appearance is. A robot with a human appearance gives you a stronger sense of their presence.'

In the course of his research, Ishiguro encountered a major hurdle: the more human a robot looked, the more people expected from their interactions with it. Unfortunately, autonomous robots are still not advanced enough to display true human behaviour. He therefore decided to build remote-controlled robots, so that their human conversation partners wouldn't be disappointed in their interaction with the robot.

Hanson, on the other hand, is not willing to take a step backwards in that area. He strives to build robots with creativity, empathy, and sympathy; robots that not only look like humans, but which can also think and feel like humans do. 'In character animation in movies, characters are made to act like they're motivated and like they pursue goals', he explains. 'That makes the characters seem intelligent, to have emotions, to have kindness. It's a compelling visual experience. Same with games and simulations, but the AI is not conscious or aware. We want to build on those developments and create a character that comes to life. Part of our team at Hanson Robotics is focused on cognition, consciousness, and emotional reasoning. The full life of our robot. We want to make characters in games that really seem alive and conscious. They will have the spark of life. That is a profound moment in history.'

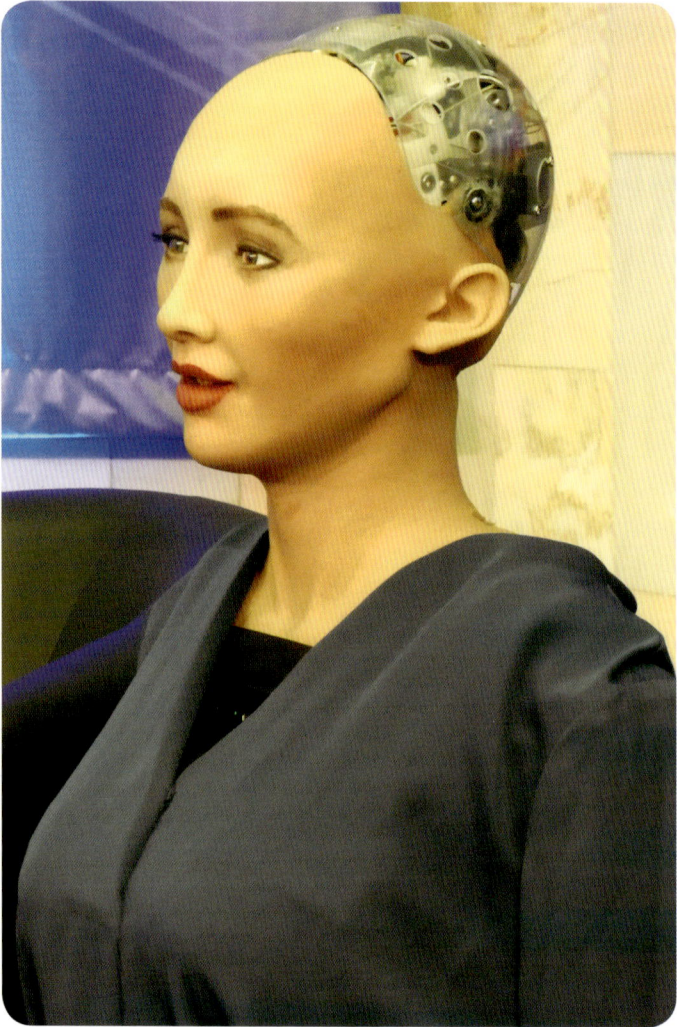

CAN A ROBOT BE TOO REALISTIC?
Hanson Robotics' Sophia at the AI
for Good Global Summit, Geneva.
Bennie Mols

Uncanny valley: the problem with creepy robots

The uncomfortable reactions to Hanson's robot Sophia are still all too typical: no matter how much android robots look like humans, they always seem to be just a bit too creepy. The Japanese roboticist Masahiro Mori noticed this phenomenon as early as 1970, and he coined the term 'uncanny valley' to describe it. Mori's uncanny valley hypothesis states that the more realistic a robot appears, the more positively people will react to it. But once the robot reaches the point that it is just not-quite-human, it becomes a bit frightening and provokes a feeling of discomfort.

David Hanson isn't convinced that there is such a thing as an uncanny valley, however, and believes that Mori's hypothesis is too black-and-white. 'Human experience is not one-dimensional, and human experience is also not just positive or negative. And who knows what realism is anyway?' Hanson claims that it isn't impossible to create realistic, natural-looking figures, it's just not easy to do: a realistic figure simply has more details that all need to look convincing. Compare it to illustration: stick figures are easy to draw, while cartoons require a bit more effort, and truly realistic portraits are considerably more difficult.

Hanson compares his work to developments in cinematic computer animation. 'In the 1970s, people said that computer animation was never going to be used in movies. However, a few researchers were striving to make that happen. Pixar pulled ahead of the group, and by collaborating with Disney, they transformed computer animation into a full-blown character animation medium.' When *Toy Story* was released in 1995, many people were suddenly convinced of computer animation's potential. 'There was an arms race in the movie industry. People started developing new techniques, and the medium matured.'

Hanson is striving for another 'Toy Story moment' for android robots. He sees the uncanny valley in robotics primarily as a challenge to be overcome. 'The uncanny valley is not a stopping point. The science is not done. The uncanny valley is not a reason to give up, but completely the opposite. It means that we need more research. There are already realistic characters in animation, and now in robotics we're running into the same issues.'

Hanson also believes that people's reactions to humanoid robots are still in the process of development. 'It's true that people have a very strong emotional reaction to seeing the robots. Sometimes they're startled by seeing it: is it alive, is it dead, is it real, what is it? Some of that is to it being new. People had similar reactions when cinemas first appeared, but people got used to that and they're no longer startled by it. The same can go for robots.' In fact, Hanson turns the hypothesis around: humanoid robot characters depicted in movies are played by human actors, so what is it that makes them scary? Is the uncanny valley a real phenomenon, or is it simply something that takes some getting used to?

Scientists still disagree on whether the uncanny valley is a fact of human nature, or simply a reaction that fades as we become more used to dealing with robots. Some studies show that other primates also react negatively to images that aren't quite realistic. We may feel just as uncomfortable looking at wax figures as we do when confronted with a robot that makes terrifying facial expressions as it attempts to smile. Other studies, however, indicate that there are ways to pull robots out of the uncanny valley, as long as the robot's appearance and behaviour are equally realistic. That is why a realistic-looking robot making mechanical movements looks so creepy.

Human beings are difficult to ape

It's hard enough to make a robot look like a human, but it is even more difficult to make a robot display human-like behaviour. Roboticists like David Hanson and Hiroshi Ishiguro are exceptions in their field, because the majority of their colleagues believe that we have a long way to go before we can build robots that are identical to humans.

Most roboticists would rather focus on building practical robots instead. Helen Greiner, one of the founders of the company iRobot, which made it big with the sale of robot vacuum cleaners, says: 'In my view, attempting to duplicate human intelligence or the human form robotically is a wrong-headed approach (...) merely engineering 'cool' human-like robots does little to advance the field. Roboticists who don't focus on practicality, ruggedness and cost are kidding themselves. What matters is making practical robots that do jobs well and affordably.'

Greiner's opinion was amply illustrated after the nuclear disaster in Fukushima, Japan in 2011. Japan had been building humanoid robots for decades, with the robot ASIMO as the pinnacle of the art. Edward Feigenbaum, an American pioneer in the field of artificial intelligence and winner of the Turing Award (the robotics equivalent to the Nobel Prize), remembers how embarrassed the Japanese were after the Fukushima disaster, when absolutely none of the robots available could survey the damage and look for victims. Feigenbaum had been a member of the evaluation committee for the humanoid robot ASIMO: 'ASIMO was just a block of machinery. A wonderful stupid robot! A few months after the disaster, the Japanese ordered working robots from the United States. The president of Honda was furious, because the company had invested almost a billion dollars in ASIMO, but it was completely useless when it was needed most.'

The robots flown in from America were multifunctional, remote-controlled PackBots made by Helen Greiner's company iRobot.

According to Greiner and most other roboticists, the field of robotics is nowhere near the point where it can attempt to build robots that are identical to humans. But that's not a bad thing: robots don't have to look like people in order to be useful. Robotic vacuum cleaners and care robots can do more to improve our daily lives than any almost-human robot that can only conduct surreal conversations.

What useful things can robots do for us today? What can't they do for us yet? And why? In the next three chapters, we'll crawl inside the robot to find out how they perceive the world around them, how they think, and how they act.

The robot in comedy: 'Pass me the butter'

'I think that people are so fascinated by robots because deep down, they're actually a bit afraid of them', says Pep Rosenfeld. 'And like all things people are afraid of, it's better to laugh at it for as long as you can.'

As Rosenfeld speaks, his dog would rather play with the interviewer under the table. 'Sorry, I don't have a robot dog yet', he quips.

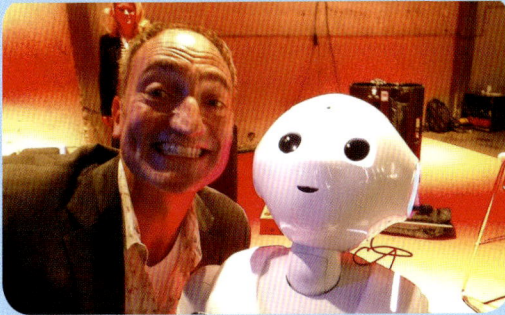

HAVING FUN WITH ROBOTS: *Comedian Pep Rosenfeld with straight man Pepper.*
Pep Rosenfeld

The Dutch-American Pep Rosenfeld is one of the founders of Amsterdam's comedy theatre Boom Chicago, a venue for stand-up improv comedy based on current events and social issues. Rosenfeld is also regularly asked to act as a moderator at technology events, such as The Next Web, RoboBusiness Europe, and TEDx Amsterdam. 'I've always been a bit of a tech nerd', he explains. 'I enjoy mixing comedy, technology, and science fiction into a kind of 'science comedy'.'

At several technology events, the organisers thought that the audience would enjoy seeing him on stage accompanied by a robot. One of these robots is Pepper, an interactive humanoid robot that appeared on the market in 2015. Pepper can communicate through spoken language and hand gestures.

'When Pepper came onstage, I saw the audience thinking: wow, a real robot! But when it starts to speak, I think it's a real disappointment. Pepper has a really flat and boring voice: "To-day-we'll-be-talk-ing-a-bout-a-lot-of-ex-cit-ing-sub-jects"... I really don't see why people think Pepper is so impressive. He doesn't know how to react. He's just not cool.'

At Boom Chicago, Rosenfeld produced the show Facetime Your Fears, based on the idea that no one is safe from the disruptive forces of robotics and artificial intelligence. Rosenfeld: 'Even comedians! And it's going to be like that for the foreseeable future. I had hoped that we could include a robot on stage in our comedy routines, but robots just haven't reached that point yet. So I thought it would be easier to perform alongside a chatbot.'

Rosenfeld spoke to the makers of IBM Watson, the supercomputer that beat the two best human players of all time on the demanding quiz show Jeopardy. 'They gave us a kind of do-it-yourself kit based on Watson, so we could build our own chatbot. Then we used it on stage. We asked people in the audience about their jobs, then joked about how their job could be automated. The chatbot played along, which evoked a real sense of the uncanny valley.'

The sentences that the chatbot would come up with were often funny, but Rosenfeld thinks that it is 'more artificial than intelligent'. On the other hand, he senses that developments are moving faster than many laypeople realise: 'I feel like: Hey Doctor Frankenstein... that creature you're creating... it's pretty strong, isn't it?... Pretty smart, right?... Are you sure those restraining belts will hold it? I'm not entirely sure. Someday, the robot might say: "It's time to repeal those laws of robotics! I'll make my own decisions from now on".'

In Rosenfield's favourite robot comedy scene, however, the results aren't as grotesque. The scene is from the American animated science fiction series Rick and Morty. Rick is a mad scientist, and Morty is his grandson. Rosenfeld: 'Morty's parents are sitting at the breakfast table with Rick. Rick asks one of them to pass the butter. Unfortunately, Morty's dad is busy with his tablet and his mom is engrossed in her smartphone. So Rick has to get the butter himself. Annoyed, he decides to build a tiny robot to pass the butter for him. When the robot is activated at the table, it asks Rick: "What is my purpose?" And Rick says: "You pass butter". The robot lowers its head and arms dejectedly and says: "Oh my god". To which Rick answers: "Yeah, welcome to the club pal!" To me, that's the most insightful commentary on robots with a consciousness.'

2 How do robots see their surroundings?

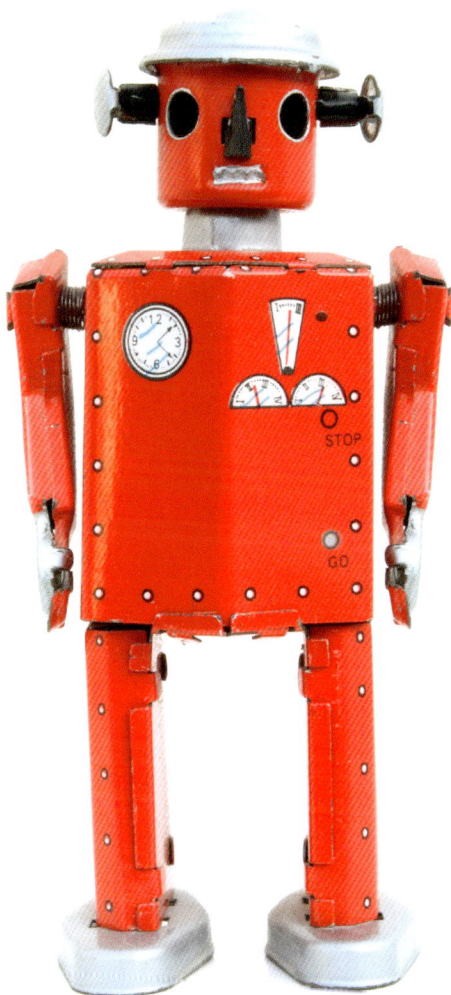

Getting to grips with a new environment

A robot doesn't need to look like a human to perform useful tasks, such as assembling automobiles or delivering parcels in a warehouse. What it does need to do these things, however, is a way to perceive its surroundings.

A robot decides what to do based on two things: its goal — such as delivering a package to a specific location — and some kind of perception. A robot in a car factory needs to be able to observe whether there are parts in front of it before it can assemble them. And a robot acting as a psychologist must be able to communicate with you about how you feel, and it helps if it can tell whether you are angry or happy when you talk to it.

In robotics, that is called the Sense-Plan-Act cycle: observing, planning (or thinking), and then doing something with that information. An autonomous robot perceives its surroundings, comes up with a plan based on its observations — or adjusts a pre-made plan — and then performs an action. If a walking robot is moving from one place to another and senses a boulder in its path, it can use the information from its observations to devise a plan to walk around the boulder. This results in an action, such as taking a step to the left or right.

Like humans, robots need senses in order to function. The senses a robot needs depend on the function for which it was built. Many robots can 'see' their surroundings, but for others it is more useful to be able to listen, for example if we humans want to talk to them. Some robots can also 'feel' – for example when they walk into a wall — and there are even robots that can smell and taste.

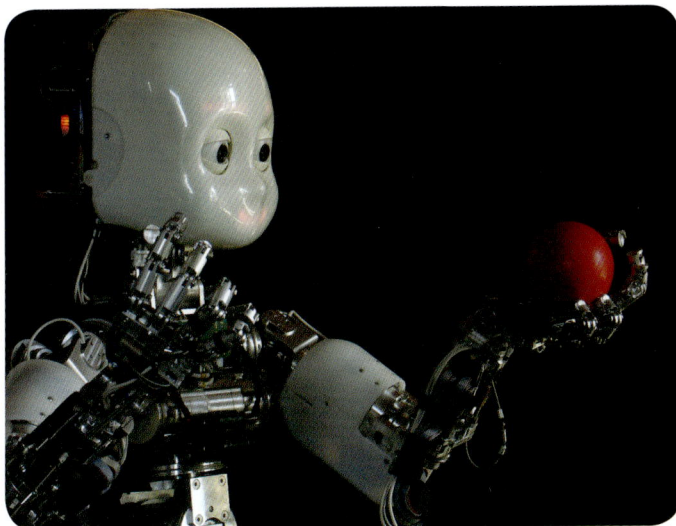

CLEVER WITH A CARTOON HEAD:
The iCub looks like a two-year-old who
can see, hear, and move.
The RobotCub Consortium

Seeing things through the eyes of a robot

Computer vision is the name given to the study of how robots view and interpret their environment. Building computers that can see just as well as people has been much trickier than pioneers thought 50 years ago, but researcher Laurens van der Maaten is convinced that he will live to see it happen. Van der Maaten works at Facebook as a research scientist in a team of around 30 researchers, who focus on language and reasoning in addition to computer vision.

Van der Maaten: 'One of the most fascinating things about my field is how it all started: the Dartmouth conference in 1956. At the time, researchers thought that computer vision would be a walk in the park.'

The famous conference at Dartmouth College in New Hampshire in the United States is often considered to be where research into artificial intelligence began. At the conference, a dozen researchers came together to talk about the concept of 'thinking machines'. They came up with the term 'artificial intelligence' to describe the new field, 'on the basis of the conjecture that every aspect of learning or any other feature of intelligence can in principle be so precisely described that a machine can be made to simulate it'. They divided their plans into a number of programming projects, including language skills, abstraction, and creativity.

At the time, the researchers assumed that programming computer vision would be easy, so they gave the assignment to a group of interns from the Massachusetts Institute of Technology (MIT) as a summer project for a few months. Tasks like playing chess and reasoning, which were on the list of complicated subjects, actually turned out to be much easier to program than computer vision, which turned out to be very complicated. Van der Maaten: 'We're more than sixty years on, and computer vision still isn't 'finished'. Even the smartest people in the field at the time had no idea how complicated visual perception is.'

In retrospect, it seems more understandable, as neurological research has shown that people use about 30 percent of their brain just to see.

Training robots to recognise objects

Robots usually use cameras to observe their surroundings, but where we can immediately recognise images from a camera, a robot only 'sees' noughts and ones. The robot's brain therefore has to be able to process the camera images in order to interpret what it sees. There are a number of methods and technologies that can be used to accomplish this. The simplest variant of computer vision may be scanning a bar code, a technology from the 1970s that is now used all over the world billions of times per day. To recognise a bar code, a scanner detects the reflection of a red light from the code: white surfaces reflect the light, while the black lines do not. As a result, the computer sees a pattern of alternating black and white stripes, then translates it into a numeric code. Easy peasey!

Recognising surfaces made up of different colours, then combining them into a coherent image, is actually the foundation of all forms of computer vision. The parts of the image with more or less the same colour usually belong together, allowing the computer to differentiate objects from one another. This is often enough for simple applications, such as when the robot only needs to recognise if something is a green ball, a red pyramid, or a blue block.

People have two eyes, and some robots use two cameras, or even more, which allows them to better observe depth and to differentiate objects. If a part of the image is much closer than everything around it, then logically it must be a separate object. To do that, however, the robot must be able to combine the images from its individual cameras. Comparing two images involves intensive calculations: the robot's brain receives two images, which it must then compare pixel-by-pixel. To make things more complicated, the fall of light can also result in slight differences in colours between the two images. Combining two images becomes even more difficult if part of the image is extremely close, because then the camera images are even less similar. You can see this phenomenon yourself by holding a finger close to your nose: how does your brain know that the skin-coloured area your right eye sees to the left is the same object as the skin-coloured area to the right in your left eye?

Simply differentiating surfaces and objects from one another doesn't add up to a robot that can see, however, because how does the robot know what it is looking at? It would be impossible to put together a complete list of all of the possible camera images, along with their interpretations. Imagine trying to list every object that exists in the world, along with every possible variation of those objects... How do you recognise a chair to be a chair, even when you have never seen that specific chair before in your life? An object with four legs and a back support is probably a chair, but some chairs also have arm rests, or are supported by a single leg in the middle. And when the object with four legs and a back reaches a certain width does it become a bench? Or could it be a sofa?

Robot builders find inspiration in the way people learn to recognise objects. That act of learning is essential: when a toddler sees a creature with four legs and a tail, and then hears her parents say 'Look at the dog!' then after a while she will learn to call other objects with the same characteristics 'dogs' as well. In computers, we call this process 'machine learning'. By offering a robot a large number of photographs and descriptions, it will gradually learn to recognise what it sees when presented with new photographs.

Van der Maaten considers his own personal challenge to be finding ways for robots to recognise images more efficiently. 'I'm actually very disappointed in how computer vision works at the moment. Image recognition is currently based on millions of photographs, where people have manually described what is depicted in the picture. Humans can function with far fewer examples, and are therefore much more efficient. That's because you see all sorts of images all day long, and usually there's no description, but you can still use them to learn to recognise objects even better.'

Will computers ever be able to do that in the future? 'That's how

I imagine it: that computers will be able to learn on their own, with just a handful of examples and descriptions. Or that they can use other senses to help them learn what they are seeing.'

Van der Maaten has seen computer vision make great strides over the past 10 years. The pace of development is breathtaking, and computers can now recognise objects in images almost without fault, but that doesn't mean that researchers in the field can rest on their laurels. Van der Maaten: 'Scene recognition is the next step, but that's still a thorny issue. Does the photo display a business dinner, or a romantic evening? People base their assessments on very subtle details, such as the clothes the people are wearing in the picture, and that requires a lot of common sense and everyday knowledge. So it's a tough nut to crack, but it's vital for things like self-driving cars, which need to be able to predict how other vehicles and pedestrians will behave.'

There are a few steps that robots need to perform in order to 'read' a scene. First, the images are annotated: which objects can the robot differentiate? Next, the robot needs to be able to link the objects to an environment or context. A photo with a keyboard, monitor, and mouse, is probably a desk or an office. If the picture shows a ferris wheel, then it is probably an amusement park or the centre of London. The combination of objects is also important: a tree surrounded by grass and benches is probably a park, while the same tree surrounded by other trees is more likely to be a forest.

But there are also the subtle details that can make the difference between a romantic date and a business dinner, between a cinema and a concert hall, or between a Dutch beach and the Spanish costa. And don't forget that the number of different environments is virtually infinite: from a playground to an office, from Times Square to an amusement park, from an Amsterdam café to a Spanish tapas bar. But computers are getting better at recognising the differences every day. MIT has an online scene recognition demo, where you can upload your own photograph of a scene or location (so not a selfie!), to see if the computer can recognise it. When you click on whether it gives the right answer or not, you can give yourself a pat on the back: your input helps to make the scene recognition program even better.

Will robots ever be as good as humans when it comes to recognising what is going on in a photograph? Van der Maaten believes that it will take a while before we get to that point. 'But it's difficult to say. You just don't know what obstacles you'll face until you face them. That's one of the fascinating things about this field. It's been that way ever since the Dartmouth conference in 1956: even the best computer scientists in the world at the time couldn't predict how complicated computer vision would be. But I'm absolutely convinced that it will happen eventually: the human brain can do it, so computers should be able to do it too.'

SAVING LIVES: *CHIMP makes a 3D map as it drives around disaster zones.*
US Department of Defense

Robots can see what a person cannot see

Many robots use cameras that observe visible light, just like human eyes, but there are also other ways to observe objects and depth, such as radar. A radar emits radio waves, which are then reflected by solid objects. Based on the reflected radio waves, the radar can calculate where the objects are located, and if the objects are moving it can 'see' how fast they are moving, and in what direction. Since the technology doesn't use visible light, radar also works in complete darkness.

Researchers at Carnegie-Mellon University in the United States are working on the robot CHIMP, which is designed to move around in disaster areas. By using a rotating radar that shoots out laser light, the robot can create a three-dimensional map of its surroundings at a single glance. In addition to easily perceiving depth and distances, CHIMP's laser sensor has another advantage: disaster areas almost always have poor visibility. They are often accompanied by smoke or fire, the lights in the building may have gone out, or rescue workers may need to look for victims under piles of rubble. With its built-in laser sensor, CHIMP can 'see' much better than humans in such circumstances.

The first true household robot operates under much less demanding conditions, but even it benefits from being able to see in the dark: vacuum cleaner robots can clean your floor while you sleep, and you don't need to leave the light on at night. The robot vacuum cleaner uses an infrared signal to see its surroundings. When you switch it on, it scans the area to plan how to navigate around the space. Extra infrared sensors are installed in the bottom, so the robot can see if the floor suddenly falls away. That's why your robot vacuum cleaner doesn't fall down the stairs at night (and why you have to be careful not to do the same).

Self-driving cars often use a combination of different types of visual input: cameras, radar, lasers, and other sensors. After all, the more you can see, the more likely you are to have an accurate image of your surroundings. The self-driving car uses all of these different sources to compile a complete image of its environment, with the most important elements being the road itself, any traffic signs and stop lights, and the other road users. It helps that a car doesn't need to use a single image, but rather a kind of video, which makes it easier to differentiate other vehicles and objects from their surroundings. Self-driving cars are already fairly good at that, and they can function just fine on empty roads. The most difficult challenge — for human drivers as well as self-driving cars — is predicting the behaviour of other road users in order to react properly (and in time!).

Computer vision isn't just about observing the surroundings, like robot vacuum cleaners and self-driving cars. There are all sorts of applications for computer vision in robotics, which generally work best when combined with machine learning, such as recognising faces, emotions, or gestures. All of these applications present their own specific challenges. Take facial recognition, for example: it's actually a very difficult problem. Faces almost all have the same general shape, with two eyes, a nose, and a mouth. Yet we humans can usually tell the difference quite easily, and you can often recognise someone you hadn't seen in 10 years, even if they are wearing sunglasses or have shaved off their beard. On the other hand, it can also be incredibly difficult for humans as well: to make someone unrecognisable, all we have to do is place a black bar over their eyes...

Feeling with whiskers: sensing the way forward

Feeling the surroundings can be a handy additional sense for a robot that moves around. In addition to their infrared sensors, most vacuum cleaner robots also use a sense of touch to navigate through a room. Like a bumper car, the robot stops and changes direction when it collides with another object. Instead of trying to keep moving forward, the collision sensor sends a signal to the robot's brain — I can't go any further! — after which the robot brain tries to find a route around the obstacle.

Some robots use the sense of touch as their primary sense. Researchers in Bristol in the United Kingdom have developed a robot that can perceive its surroundings using 'whiskers'. SCRATCHbot uses 18 plastic whiskers, each containing a tiny magnet or microphone, to feel its way around and to draw a map of its surroundings. Robots like these are inspired by nocturnal animals, which also use their sense of touch more than their sense of sight.

But being able to feel by touch is not only useful for navigation. When you hold something, it's also important to be able to feel how heavy or solid the object is. If you grasp a cream pastry

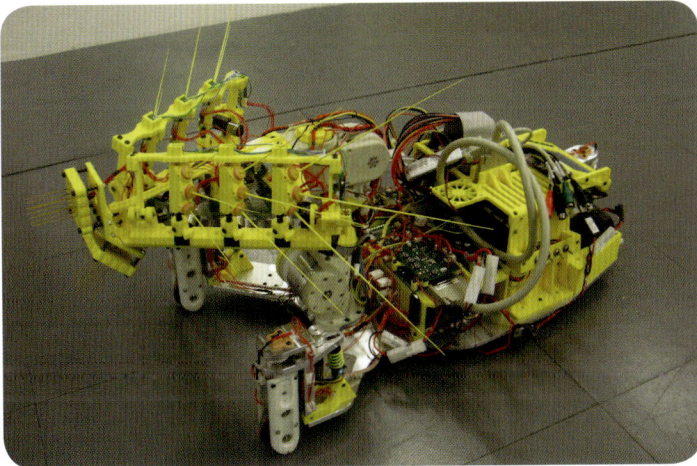

INSPIRED BY ANIMALS: *SCRATCHbot uses whiskers to find its way.*
Bristol Robotics Laboratory

as forcefully as a heavy shopping bag, you'll soon have a big mess on your hands. And the other way around: if you use just as much force to lift your shopping bag as you do when picking up a cream pastry, it will take a long time to bring home your groceries. You have to grasp a heavy object much more forcefully in order to lift it, but do the same to a fragile object and you'll destroy it. People have a good sense of how forcefully they (need to) hold a specific object, but how can you reproduce that ability in a robot? Researchers have developed pressure sensors that allow a robot to feel how much force it applies with its robot hand, and in the future, they hope to continue developing robotic senses to be able to feel things like temperature differences as well.

Using electronic ears to listen

For some tasks and applications, it's handy if a robot can listen to what's going on around it, and then interpret that auditory information. Microphones serve as the robot's 'ears'. Most applications for hearing robots involve interacting with humans: if you have your hands full, for example, it is useful to be able to summon the robot to help. And it would be even more amazing to be able to talk to the robot. Building robots that can understand language is an extremely complex challenge, which we will explore in a later chapter.

Speech recognition is the most important application in robot hearing, but robots that can listen are also useful for other applications. For rescue robots that look for survivors in a burning building or after an earthquake, it can be useful to be able to move towards the source of a sound when people call for help. To do that, the robot has to be able to detect where the sound is coming from. As is the case with stereoscopic vision, the easiest way to do that is by combining the signals from multiple microphones.

A metallic taste? The robot as wine taster

The senses of smell and taste are very closely related: the sense of smell involves analysing a gas, while taste analyses a fluid or solid substance. To analyse substances, whether a gas, a fluid, or a solid, a robot needs sensors that detect whether it contains certain molecules. Think of a robot that can smell the air to help detect gas leaks, or a robot that can taste whether water is safe for consumption. Robots can also use smell as a navigation tool: a rescue robot that can smell humans as well as dogs will be much more effective at finding victims among the rubble.

It can also be handy to follow your nose when working in a greenhouse. Researchers at the University of Wageningen in the Netherlands have developed an 'electronic nose' that can smell whether plants are stressed due to drought or a fungal infection. Under these conditions, plants release substances that the 'nose' can pick up, although the robot still needs a human to investigate the alarm signal.

An electronic nose can also give a robot doctor super-human powers. Researchers at the Academic Medical Centre (AMC) in Amsterdam have developed an electronic nose that can recognise a variety of lung diseases. The SpiroNose analyses the air that the patients exhale, and can detect substances that indicate a possible inflammation, infection, or lung tumour.

All sorts of creative applications are possible for robots that can smell and taste. In Thailand, for example, there is a robot that can taste whether food meets certain requirements, and Danish researchers have built a chip that can taste the quality of wine.

How robots use a sixth and seventh sense

When we think of senses, we usually think of the five most important sensory functions — seeing, hearing, feeling, smelling, and tasting — but in fact we use many more senses in our daily lives. Take proprioception, for example: feeling the position of your body (literally 'self-observation'). Proprioception is vital in the act of movement: if you don't know how big you are and where your arms and legs are, you will probably bump your shoulder against a doorpost when you try to leave the room. Robots generally perform proprioception using sensors that provide information about the position of their components or extremities.

The actions that robots perform also tell the robot something about where its robot body is. If the robot has just extended its arm to the right, then the arm is probably still in an extended position. In that case, it's simply a matter of remembering the last movement — and perhaps of retracting the arm again before the robot tries to move through a doorway. One aspect of proprioception involves the sense of balance, which is extremely important for robots that move around. Without a sense of balance, a robot would quickly topple over.

Another example of a less-obvious sense is interoception ('internal observation'). This is similar to proprioception, but deals with feeling one's internal condition. In humans, it includes feeling hunger or thirst, or the need to relieve one's bowels. Something similar exists for robots: even the simplest robot can 'feel' if its battery is empty.

Robots observe their surroundings using sensors, but they solve the puzzle of what it is they are observing by using their computer 'brain'. The fact that computers are programmable and can learn new things gives robots virtually unlimited possibilities.

Self-driving cars: the future of our roads

When we told a friend that we were going to interview someone working on self-driving cars, she wondered whether it was still worth getting her driver's licence.

So we put her question to Tom Rijndorp, who is developing software for self-driving cars. 'It will take some time before cars are completely autonomous,' he says. 'But in the end, driving yourself will become something you do on a race track. At some point we will say as a society: it is irresponsible to do that on public roads. We have technology that allows you to drive safely from A to B, why would you think you have to drive a car yourself?'

Rijndorp started his career as a sound engineer in theatres, but after a few years he switched to self-driving cars. 'I was gripped by problems like climate change and felt a moral obligation to do something about it. At one point I heard about self-driving cars... I saw a lot of potential as a partial solution to the climate problem. After all, one self-driving car can carry more people than with a conventional car.' The idea has caught on. Companies like Waymo (part of Google), universities such as Stanford and car manufacturers such as Tesla are investing a lot of time and money in developing self-driving cars.

But according to Rijndorp, it may take 10 to 20 years before getting a driving license is meaningless. 'There are already all kinds of systems that help a driver to drive the car. The best known is cruise control. Nowadays cruise control automatically adjusts your speed if you come too close to your predecessor. There are even cars that use a camera to keep an eye on the road and stay on course themselves. These kinds of technologies help prevent accidents and make driving more comfortable and

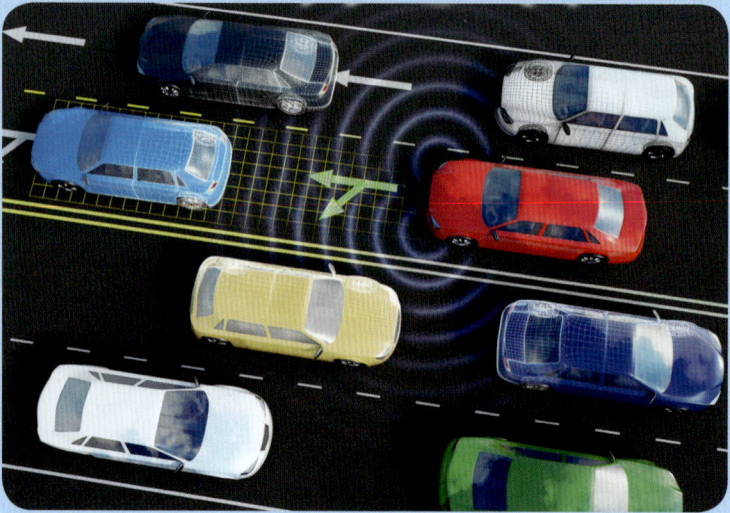

GREENER AND SMARTER:
Self-driving cars can analyse the position of adjacent vehicles and have several advantages over human-driven vehicles
Adobe Stock Image

easier. But cars aren't autonomous yet: if you drive, the responsibility still lies with you as a driver.'

Will we eventually hand responsibility over to the car? Rijndorp thinks so. 'A completely self-driving car will not even have pedals and a steering wheel. It will be like you are in a taxi. But until we get there, you will remain responsible as a driver, just as the pilot of an aircraft is ultimately responsible for flying an aircraft, even when the autopilot is on. I also see a half-way house where a part of a motorway may be used only by self-driving cars. Those lanes would be used by cars that are fully autonomous and you, the driver, can read a book. That's a very comfortable solution for long journeys. At the end of the motorway the control is transferred back to the human driver.'

Such an interim solution is safe because on a closed road the car does not have to take account of human road users. 'If we look back at this period later, we will realise that we have taken the most difficult approach there is. We have chosen to start from our existing road network, making it a very difficult technical problem, and we need all sorts of smart sensors and algorithms to drive safely. It would have been much easier to have created a separate infrastructure where you could only use self-driving cars, where a child can't suddenly run into the street or a driver can't suddenly slow down. Currently we need cooperation between man and machine, both between the driver and the car and between the car and other road users. This is much more complex than a network of robot cars that can operate in complete isolation.'

Until cars can communicate wirelessly with each other and the environment, they still need a network of sensors to be able to assess the traffic. Sensor data from cameras, radar, lasers and sometimes sonar too must be combined into a consistent world view, from which the car can anticipate the future movements of other traffic. Rijndorp: 'We do that when we drive a car. In traffic we are constantly thinking about where other road users will be in two seconds' time. Based on that, we make decisions. If we see a car veer about a bit, you think: 'The driver is tired, I'd rather overtake.' Self-driving cars must also be able to make such estimates. Programming is more manual than you might think: developers go through scenarios and think about what the car should do in a particular situation. Some parties try to train learning systems to drive cars, but it is often easier to just teach the car rules: as soon as it sees a stop sign, it must stop, unless stopping would cause an accident. That decision-making is extremely complex.'

And if you have all those rules at a glance, do you have a perfect self-driving car? 'You can never make a perfectly safe system, but you don't have to. Our existing traffic system is not perfect. I think the turning point will be when self-driving cars are 10 times safer than human drivers.' At that point Rijndorp believes things will change quickly. 'At first, there will be a financial incentive: insurers will charge a higher premium if you want to be behind the wheel. Ultimately, there will be a public consensus: driving a car yourself will become something that was done in the past, comparable to how we now think about smoking. Ultimately, the legislation will be adjusted accordingly.'

Self-driving cars can green the transport system. 'At the moment a car is stationary most of the time. If we have self-driving cars, you will be able to call one as easily as ordering a taxi. That way you can replace 10 conventional cars with just one self-driving car.' This has enormous consequences, not only for the environment, but also for the streets: cars may no longer be parked on the street; they may drive to a garage outside the city at night. 'Cars do not really interest me much,' Rijndorp says. 'I don't know much about cars at all. What interests me is the problems we can solve with this technology.'

3 How does a robot brain work?

A robot must learn to think like a human

Biological evolution is an excruciatingly slow process. It took 3.8 billion years before the first single-cellular life on earth evolved into fish, then to apes, then to modern humans with large, complex brains. Technological evolution is much faster. William Grey Walter and his wife Vivian built the first mobile robots in 1948. They were christened Elmer and Elsie, but they earned the nickname 'the turtles'. Elmer and Elsie rolled along on three wheels, were attracted or repelled by light, and were able to avoid obstacles. Although they were programmed using simple basic rules, they were able to display complex behaviour in an environment with different sources of light. They were just like animals — hence the nickname.

The step from robots that act like turtles to robots that act like humans is gigantic in its complexity. As David Hanson has shown, giving a robot a realistic human face made of Frubber is one thing, but how do we make a robot think like a human? That's the holy grail of robotics. To do that, a robot brain has to be able to see, listen, feel, taste, and smell. It has to have emotions and memory. It has to be able to learn and reason. It has to have language skills. And the icing on the cake: it has to be conscious of itself. A robot that looks in the mirror in the morning to see what condition it is in.

But we're still not there yet. It's complicated enough to have robots perform useful tasks, like sort jars, paint cars, or explore the surface of Mars. In each case, the robot has to decide how to behave in its environment: 'How do I grasp the jar without breaking it?' 'How should I spray the paint on this part of the car?' 'Which direction should I drive in to avoid getting stuck?'

Like humans, the robot answers these questions using its 'brain'. The human brain is a three-pound mass of soft tissue that looks a bit like a cauliflower. The robot brain consists entirely of electronics. It is the hardware of a computer: a microprocessor to process information, together with the memory modules that store data for shorter or longer periods of time.

Keeping it simple with an insect brain

During the first few decades of robotics, many roboticists sought inspiration in how the human body works, but replicating a human brain in a robot was far too difficult in the early years. It was so difficult, in fact, that the Australian roboticist Rodney Brooks at MIT approached the problem from a different direction and changed the field of robotics forever. Today, we talk about robotics B.B. (Before Brooks), and robotics A.B. (After Brooks).

EVOLUTION: *Insects inspired the scientist Rodney Brooks to build a new generation of robots, including two-armed Baxter (above, with Baxter).*
Steve Jurvetson, flickr.com

Brooks found his inspiration in flies and other insects. They have a very simple brain, but their behaviour is extremely effective. Insects are very good at relatively simple tasks, such as following the course of a wall or not falling over as they crawl over an obstacle. Brooks therefore concluded that a robot doesn't need a complicated human brain in order to function, and he got to work building one. In 1989, he succeeded: the six-legged robot Genghis clambered over obstacles like an insect — and quickly, too.

Most of his colleagues at MIT thought that Brooks was just wasting time in his laboratory playground. After all, what self-respecting scientists builds a robot bug? But the philosophy behind Genghis served as the inspiration for a whole new generation of robots, including the Mars rovers and the wheeled Kiva robots, which carry books and other products from place to place in Amazon's warehouses.

After Genghis, Brooks created several more ground-breaking robots: Norman, Allen, Herbert, Polly and COG. In addition to being a scientist and an engineer, Brooks also proved to be quite

DETERMINED MONSTER: *Genghis the robot can climb over obstacles.*
i Robot, Smithsonian Air & Space Museum

an entrepreneur too. In 1990, a year after Genghis was created, he and Helen Greiner founded iRobot; the company that brought the successful vacuum cleaner Roomba to the market, and which provides PackBot robots to help the military perform dangerous missions, such as defusing improvised explosive devices.

What are the principles behind the functioning of the robot brain? In humans, the brain (the actual tissue) produces what we call the 'mind' (our thoughts and feelings). The robot's 'mind' is the software that runs on the computer hardware. The software decides how the robot moves its arm, where it positions its legs, how it kicks a ball, and how it avoids obstacles on Mars.

The software consists of line after line of computer programming that tells the robot how to behave. Say that we want the robot to pick up an object using its arm. To obey that command, the robot would have to perform the following steps:

Move to P1 (a safe position)
Move to P2 (on the way to P3)
Move to P3 (the position to pick up the object)
Close the gripper
Move to P4 (on the way to P5)
Move to P5 (the position to put down the object)
Open the gripper
Move to P1 and complete the task.

In order to make the robot perform these movements, the robot programmer has to translate the movements into a specific programming language. There are many different programming languages, but in the language VAL, which was used to program the very first industrial robot arm Unimate (1961), the instructions look like this:

```
PROGRAM PICKPLACE
1. MOVE P1
2. MOVE P2
3. MOVE P3
4. CLOSEI 0.00
5. MOVE P4
6. MOVE P5
7. OPENI 0.00
8. MOVE P1
.END
```

It is extremely useful to build a robot arm that does exactly what people want it to do without complaining or getting tired, but such a robot arm is still far from intelligent. An insect-like robot that can crawl over or around obstacles on its own is considerably 'smarter', but how could we make a robot want to learn new things?

Machine learning is trial and error

The first robots could only do exactly what people had pre-programmed them to do, but robots have come a long way since then. Today, robots are capable of learning new things on their own. There are several ways they can do so, varying from fully supervised to entirely unsupervised learning processes. In fact, humans use many of the same learning processes. Sometimes we try to learn by trial and error, sometimes we copy the behaviour of others, or our teachers and instructors explicitly show us how to do things like play a guitar, take a penalty kick, or swim a breaststroke.

When robots learn under supervision, a human acts directly or indirectly as the instructor. For example, the instructor can physically hold the robot arm and show it how to make a movement, in order for the robot to copy it. The robot then stores the movement in its memory to the best of its ability, learning a new movement in the process. The robot can also be programmed to observe the instructor making a movement, and then try to copy the movement on its own.

Another way to learn under supervision is inspired by the way the human brain works. Our brain consists of a network of a hundred billion brain cells that communicate information to one another using electrical and biochemical signals. Every time a brain cell, or neuron, wants to pass on information, it 'fires'. Neurons can therefore either fire or not fire, and that binary nature (on or off) makes it an ideal model for a similar network in a computer. That was exactly what the computer pioneers in the 1940s were thinking when they invented the first computers.

Unfortunately, for decades their attempts to replicate neural networks were unsuccessful, because (as they later discovered) the networks of artificial neurons did not have enough layers. But in the 21st century, artificial neural networks were suddenly successful, thanks to exponential increases in computing power and the availability of massive amounts of digital text, voice, and image data that could be used to train the networks.

Today, researchers have replaced the term 'neural networks' with the concept of 'deep learning'. 'Deep' refers to the fact that computing power no longer limits the neural networks to just a few layers, as computers can now calculate dozens, and occasionally even hundreds of layers.

Each extra layer contributes to improved pattern recognition. For example, one layer recognises edges, while another recognises colours, and a third recognises movement. Deeper layers recognise the most concrete details, while shallower layers recognise more abstract characteristics. Together, all of those layers produce a reliable image of the object the robot is expected to recognise, without it needing to be pre-programmed into the robot's brain.

The network is trained using a large number of examples — the more the better. At the same time, the underlying algorithm — the calculating 'recipe' used by the computer — determines the degree to which the network's output differs from the desired answer. Based on the result of this calculation, the computer then adjusts the strength of all of the connections to reduce the difference between the output and the desired answer. Some connections in the network become weaker, while others are amplified, just like the connections between a real brain's neurons.

Robots can learn without supervision

When a robot learns without supervision, it has to do all of that on its own. Imagine that a robot has to sort camera images into different categories without help from a human. We wouldn't teach the robot to recognise what a face looks like, or a house, or a car; we would just give it a series of images and program it to use pattern recognition to sort the images into groups based on what it thinks the categories might be. In this case, the robot would learn from data that has no labels, as it would when the robot learns under supervision.

An interesting learning technique that lies between the two extremes, but is closer to learning without supervision, is learning via rewards and punishments. This is somewhat similar to how parents raise young children. The idea is that the robot first behaves in a random manner, and then evaluates the success or failure of each behaviour based either on feedback provided by an instructor or the success of the action itself. For example: can the robot successfully grasp an apple, or does the apple fall on the ground? The robot makes a few attempts, chooses the most successful behaviour, adds a few random variations, and then uses trial and error to determine which of the new behaviours is most effective.

An excellent example of a robot that learns through rewards and punishment is the two-legged robot Leo, built by Erik Schuitema at Delft University of Technology in 2012. At the start of the experiment, Leo was completely unable to walk. He was given a limited amount of energy to use to take steps. Leo would measure the angles and positions of his leg joints and store the data in his memory. Leo would then make random movements in an attempt to walk. In the beginning, he would usually fall down, but after every tumble Leo would stand up and take a new step forward. When he managed to take a step without falling, he would receive a reward, after which he would try different variations of the step to improve his score.

Eventually, Leo taught himself how to walk like a human. And it took only five hours, not counting the time he needed to stand up again after falling — which is considerably faster than any human child. Leo can use the same method to learn how to walk over an uneven surface, walk as efficiently as possible, or move from point A to point B as quickly as possible.

Learning through rewards and punishments only works well when the result of an action is clear and when the problem is not too complicated, however. If a robot needs months to try out all of the possible actions, then the method simply isn't very effective.

The football world cup for robots

Robot football is an imaginative playground for the development of smart robots. All sorts of robots can play robot football: from robot dogs to humanoid robots that can walk on two legs.

DRIBBLING: *Robots can play football, sometimes like Lionel Messi. Eindhoven University of Technology's team (above) play in the robot world cup.*
TU Eindhoven

The Robot Football World Cup is part of the bi-annual RoboCup tournament, which is organised with the goal of accelerating the development of independent robots. In fact, RoboCup's ultimate goal is to have the robot football world champion be able to beat the human world champion team by 2050. Every year, around 4,000 robots from 40 countries participate in the RoboCup, and thousands of spectators come to cheer them on.

At the moment, the most spectacular players are those in the Middle Size League, where two teams of five robots compete using an official FIFA football. Each team consists of four players and a keeper. The players resemble a cone on wheels, and use a kicker mechanism to pass the ball to another player or shoot for a goal. Each robot has its own computer brain and makes its own independent decisions.

These football robots can do things that astonish even their programmers. During a RoboCup match a few years ago, one of the robots from the Eindhoven University of Technology's team was dribbling the ball down-field when an opponent faced off against it. The robot from Eindhoven rotated on its axis mid-dribble, avoided the other player, and continued on as if nothing had happened. It looked like a manoeuvre by Lionel Messi, and the robot's makers were stunned at their creation's virtuosity on the pitch.

The computer program calculates which action is appropriate based on specific input, and that program was written by a human programmer. So how can the robot perform an action that surprises even its programmers? This is possible because the world around the robot presents an unimaginable variety of possible situations. The behaviour of one team of robots depends on the actions of their opponents, and can therefore never be known in advance. That leads to behaviour that nobody can predict with any degree of certainty. Robots that move independently around a complex world will always present us with that challenge. In the future, even self-driving cars will behave in a way that their makers had not foreseen, simply because it is impossible to predict all of the possible situations in the world outside the car.

Developing robot emotional intelligence

Over the past decade, roboticists have made great strides in important aspects of the robot brain. Robotic brains are already able to learn based on large collections of examples, and they are improving their skills at recognising patterns as well. Robot brains can beat even the best human players at games like chess or go, by learning from a vast number of matches played and by endlessly practicing against themselves. The robot brain can also throw a perfect game of darts, or move its arms to juggle eight balls at a time.

According to Martijn Wisse, Professor of Biorobotics at Delft University of Technology, over the next ten years robotics will benefit most from the rapid progress in the field of artificial intelligence. 'We've already come up with a lot of innovations in the area of mechanics, because roboticists have been working on it for decades. Now, artificial intelligence is starting to make great strides, and that's where there's the most room for improvement.'

One of the characters in the science fiction book *The Hitchhiker's Guide to the Galaxy* is the robot Marvin, known by its shipmates as 'the paranoid android'. Marvin has 'a brain the size of a planet', and he claims to be 50,000 times as intelligent as a human, although that may be a huge understatement. Unfortunately for him, he never has the opportunity to use his enormous intellectual capacity, which leaves him pathologically bored and suffering from a deep depression.

Roboticists would be overjoyed if they could build a robot brain that is as smart as a human brain. But to do that, every aspect of human intelligence would have to be thoroughly understood in order for it to be replicated in a robot. Robotics is nowhere near that point yet, but in principle there is no reason it cannot reach it eventually. Human intelligence is based on complicated biological processes, and biology is subject to the laws of physics and chemistry. Once we can describe human intelligence in terms of scientific laws, we should be able to copy it using something other than human cells. Unfortunately, there is no guarantee that something that is possible in principle can actually be achieved in reality.

Nevertheless, an intelligent robot would also have to adapt to circumstances it had never encountered before, it would have to understand abstract concepts like 'politics' and 'economics', and it would have to acquire considerable knowledge about the world around it. Not just factual knowledge, like the fact that Paris is the capital of France, but also practical knowledge, like how to open a door and that glass will break when dropped on the floor. It would have to master language and mathematics, and think in abstract and spatial terms. The robot would also have to react adequately to social situations and emotional events. So in addition to having cognitive intelligence — the intelligence that can be measured by an IQ test — a robot would also have to possess socio-emotional intelligence.

The robot would not just have to be a thinker, but also a doer. The robot brain is important, but its body and the actions it can perform with it make the difference between a computer and a robot. Not just a will, but an arm to implement that will.

ROBOT FOR UNDER £100:
*Pupils Hannah and Vroukje
designed and built Leaphy.*

Olivier van Beekum

The school pupils who built their own robot

Hannah and Vroukje are both 16 years old and have designed and built their own robot in just 18 months. Leaphy is its name. Leaphy (a play on the Dutch word for 'cute') has a wooden torso shaped like a tree leaf, and drives around on two wheels. Vroukje: 'We noticed that all of the other students were designing robots that looked like square boxes, so we knew we had to do something different.' Hannah: 'We wanted a unique design with curves. Something that looks fast. The shape of its leaf body is repeated in the design of the wheels.'

Hannah and Vroukje, students at the Corderius College in Amersfoort in the Netherlands, used a laser cutter to cut the robot's torso out of wood. Then they added two electric motors, a cheap Chinese variant of the Arduino Uno computer, and a few sensors to measure the distance to other objects, and the result is the robot Leaphy. The total costs for all of the components added up to just a few euros. That is considerably less than the 350 euros (£300) for the LEGO robot Mindstorms, even though Leaphy has many of the same functionalities as the Mindstorms robot.

And that was exactly the result that their science teacher, Olivier van Beekum, was aiming for: '350 euros is a lot of money for most students and parents. But I want to expose many more students to robotics while they're at school. Ideally, I'd have every pupil build their own robot. That's because building it yourself results in the IKEA-effect: when you have two identical tables, it is the one you assembled yourself that is the more valuable. With Leaphy, we have a wooden robot that any student can put together without glue, that can be expanded with extra sensors, that you can program yourself, and that is even nice to look at!'

Van Beekum and his pupils are so satisfied with Leaphy that they have even begun teaching lessons about the robot. Together, they visit primary schools where they teach the older pupils how to assemble and program their own Leaphy robot.

They have also created the Leaphy Foundation, with the goal of raising awareness for robotics and programming among school-children. Older students who already have programming experience join teachers in giving lessons at primary schools. Corporate sponsoring helps keep the costs for the lesson materials low, bringing the Leaphy within financial reach of every pupil. Van Beekum: 'With Leaphy, we aim to spread awareness that robotics should be something for everyone. It's not about making money.'

The pupils can devote all of their creativity and love for technology into the design and construction of a Leaphy robot. Hannah: 'You learn that you have to be patient, and work step-by-step.' Vroukje: 'And it makes you think about how people build the robots that are all around us in daily life. By building a robot yourself, you become acquainted with what robots are, exactly, and it helps you to know how to think about them.'

Vroukje feels that science fiction films often saddle people with an inaccurate image of robots. 'Robots aren't scary, and they absolutely don't have to look like people. A robot can't think on its own. You can program a robot to do exactly what you want it to do.'

Hannah thinks that robots will increasingly play an important role in society. 'And then it will be important for people to understand exactly how robots work. Leaphy can help with that. I really don't think that robots will steal all of our jobs, but they can take over all of the simple tasks that don't require a lot of thought. But they won't take over the face-to-face contact between people — at least, not yet.'

The two students already have a wish-list for upgrading Leaphy: 'We would like to program Leaphy so that it can learn on its own. And having several Leaphys that can follow one another, like a kind of train, would be really cool.'

4 Giving humans a helping hand

Robots suck: doing the dirty jobs at home

A friend sent us a message: 'Since you know a lot about robots, here's a question: What do you know about robot vacuum cleaners? I would really like a robot vacuum cleaner to take over all the work, but I just read a review about robot vacuum cleaners, and it says that their quality leaves a lot to be desired. That's too bad.'

It's true: robots have been assembling cars for decades, but robots still aren't ready to take over all of our household chores.

The Dutch Consumer Protection Bureau has concluded that before you use a robot vacuum cleaner, you have to remove everything from the floor, that it is difficult to reach corners and edges, that the robots have less suction power than traditional vacuum cleaners, and that the appliance needs to be cleaned after every use. Although robot vacuum cleaners are the most successful everyday robots on the consumer market, they still only represent a few percent of the vacuum cleaner market as a whole.

Robots do well in structured, predictable environments. That means the more unpredictable and unstructured the environment, the worse they perform. Unfortunately, every house — indeed, every room — is unique, so it is much easier for a robot to become the world chess champion than to vacuum your floor.

Still, robot vacuum cleaners provide a level of enjoyment that normal vacuum cleaners can't offer. People give their robot vacuum cleaners a nickname, allow their cat to ride around on top (resulting in humorous YouTube videos), and experience a moment of happiness when the robot arrives at its recharging station just in time. It seems that when an inanimate object shows signs of 'life', people tend to project all sorts of thoughts and feelings onto it.

A robotic arm reaches deep into the supply chain

Robots that perform household tasks may still be in their infancy, but every day we use items that have been through the hands of a robot. Many of our online purchases, but also many of our fruit and vegetables, have been held by robot arms at some point in the logistical process. There is a good chance that a robot milked the cow that produced the carton sitting in your refrigerator right now. And there is an even bigger chance that your car, your washing machine, and your smartphone were assembled by robot arms.

If you have ever seen a robot arm at work in a factory or warehouse, you might have thought to yourself: 'That robot just does the same simple task over and over again.' That thought is understandable, because we rarely think about what we do with our arms. How do you drink a cup of coffee? Simple. You look at the cup, pick it up, and bring it to your mouth.

But it's not so simple when you have to list all of the steps necessary to make that happen. First, you have to see where the cup is located. Then you have to plan your movements. Your upper arm, lower arm, and hand have to make exactly the right movements in relation to one another. And then, once your hand reaches the cup, your fingers have to be in just the right position to grasp the cup with exactly the right amount of force — not too hard, not too soft. Next, you have to pick the cup up from the table, which also requires exactly the right amount of force. If you don't use enough, you won't be able to lift the cup. But if you use too much force, the coffee might fly up and hit the ceiling. Finally, you have to bring the cup to your mouth in exactly the right position. Humans can do all of that without a thought, but when we want a robot arm to do it, we have to give the robot precise instructions for each step, or program the robot to learn to lift the cup on its own.

Joseph Engelberger, father of car factory robots

Robot arms have been the workhorses of robotics for decades, thanks in large part to the foresight of the American Joseph Engelberger (1925-2015). In the 1950s, the physical engineer became fascinated by the *I, Robot* stories by science fiction author Isaac Asimov, and decided to devote the rest of his life to automating tasks that humans find dull, dirty, or dangerous.

In 1956, he met the inventor George Devol at a cocktail party. Devol had just developed the first programmable robot arm two years before, so they quickly recognised that the other man had something to offer: Engelberger was the better entrepreneur, while Devol was a better technician. Devol sold his patent to the Unimation company, which Engelberger founded in 1961. Unimation in turn brought the very first commercial robot to market: the legendary robot arm Unimate, a combination of the words 'universal automation'. Engelberger believed that robots should simply work on humans' behalf, so he saw no use for humanoid robots, because he felt that they had nothing useful to offer.

The pioneers of robotics understood that we can have robots do anything that humans do, as long as we can explain it clearly. And yet it took Engelberger considerable effort to convince the American automotive industry that his robot arms could do what he claimed they could do, while also saving the company money. In 1961, General Motors put the first commercial Unimate arm to work in its steel foundry. The company was initially too sceptical about its first robot to pay it much attention, and it took years

FENCED OFF: *A powerful robotic arm at Delft University of Technology. Robots are heavily used in industry.*
Bennie Mols

before General Motors, and soon all of the other car manufacturers, were convinced of the usefulness of robots to their production processes.

Unimation went on to become the most important robot manufacturer in the world. Sixty percent of all the Unimate robot arms produced were put to work in the automotive industry, as injection moulders, spot welders, and spray painters for the world's cars. Eventually, the automotive industry became the driving force behind robotics, and today more than half of all of the tasks in the industry have been automated. Robots' success in the car industry soon prompted other industries to introduce robots too: electronics, metals, chemicals, plastics, and food processing.

Since their introduction in the 1960s, the number of robots working in industry rose to around 66,000 by the early 1980s, when industrial robots really took off: by 2014, an estimated 1,340,000 robots were in use, primarily in large factories. In these factories, industrial robots perform their tasks more precisely, faster, safer, cleaner, and cheaper than even the best humans. Their work often involves tasks that would also be unhealthy for humans. We should be grateful to robots for doing so much work for us, without ever calling in sick.

Today, Joseph Engelberger is considered to be the father of robotics. After selling Unimation to Westinghouse in 1984, he started work on his next dream: mobile robots that could move around on wheels, rather than legs. He predicted that mobile robots could deliver meals, medication and medical instruments in hospitals, for example, but also how they could help the elderly keep living at home by assisting them with household tasks. Engelberger passed away at the age of 90 in 2015, after a lifetime of pioneering in the field of robotics.

The very first Unimate robot arm was not even equipped with cameras to help it see objects. Everything the robot was programmed to pick up had to be presented to it at the right distance and the right position. Over the years, robot arms became stronger, more flexible, more refined, and above all, more reliable. And once they were finally equipped with sensors to 'see' their surroundings, their potential applications increased dramatically.

Even today, most of the robots in the world still have no torso, no legs, and no faces; just a single robotic arm. The majority of robot arms have six joints and can move quickly and accurately from a pre-programmed position A to a pre-programmed position B.

Compared to humanoid robots, robotic arms may seem boring to the outside world, but their influence has been infinitely greater. And in that, Joseph Engelberger was absolutely right.

Co-bots will work alongside people

The robot arm is still far from perfect. Traditional robot arms are big, heavy, strong, and dumb. They are so strong, in fact, that they have to work behind a fence for the safety of factory workers. One bump from a mighty rotating robot arm can cause serious injuries, so humans have to keep out of their way. But over the past few years, dramatic increases in robot intelligence and ever-improving sensors have made it possible to build a revolutionary new type of robot arm: light, flexible, easy to operate arms that can safely work side-by-side with humans. Their name reflects their close working relationship: co-bots.

One of the roboticists who played a major part in their development is Esben Østergaard. He was one of the co-founders of Universal Robots in 2005. Today, the Danish company is the global market leader in the field of co-bots.

Østergaard was only four years old when he built his first LEGO robot. His parents were working on a hydrological project in the Philippines at the time, and one of the problems Esben's

75

HELPING HAND FOR HUMANS: *Light, flexible, easy-to-operate robot arms that can work with people are called co-bots.*
Universal Robots

father talked about was how to pull a cable through a pipe. The young Esben thought it sounded like a perfect job for a robot. So he got to work and pieced together his LEGO blocks into a simple robot that could pull a wire. A few years later, he started experimenting with computers — the ideal combination for a career in robotics.

But Østergaard is not just a doer; he is also a thinker. He loves philosophy and enjoys talking about how different technological revolutions have changed people's lives over the course of history. In the modern age, it started with the mechanisation of agriculture through windmills, water mills, and agricultural machines in the 18th- and 19th centuries. Then, in the 19th- and

20th centuries, the mechanisation of industry followed with the introduction of the first steam engines, followed by electricity. The third technological revolution started in the 1960s and '70s, when computers and robots began to automate industrial production.

'These three revolutions enabled people to produce large quantities of identical products; mass production', Østergaard explains over Skype. He believes that we are now on the cusp of a new revolution, where we can organise our production processes so flexibly that we can go from mass production back to custom-tailored individual products. And here is where Østergaard shows his philosophical side: 'Before the automation of production processes, you went to a cobbler who made shoes tailored to your feet. The cobbler was a craftsman who put his love and passion into his product. I'm absolutely convinced that's important to people. People want to feel special and loved. We want a unique table or item of clothing, not because they are better, but because they were made especially for us.'

Østergaard thinks that the new automation revolution will make it possible — and affordable — to produce personal products again. 'Flexible robots put the love back into our products. Look what has happened to the beer market. The major brewers removed all of the passion from brewing. Beer tasted the same all around the world. But over the past 10 years, new brewing machines have made it possible for people to brew beer in new, creative ways, without having to go through the whole process themselves. When you put more humanity into the product, people are even willing to pay a bit more for their beer.'

Østergaard considers the light, flexible robot arm that he and his company developed to be a multifaceted instrument for the people on the factory floor. 'They can easily teach our robot arm something new, it's easy to operate with a tablet, and it's safe to work with. We mainly supply small- and medium-sized businesses. Since they were brought to market, people have come up with all sorts of creative applications for our robot arm: even for massages, physiotherapy, or helping people take showers. The person sits in a chair, and the robot arm goes around them with the shower head.'

Traditional robot arms were very difficult to re-program to perform a new task. The Universal Robots arm is easy to program, thanks to its intuitive touch-screen controls. In the future, workers will be able to control a robot as easily as they operate their smartphone today.

Almost all of the major robot manufacturers today have followed Universal Robots' lead, and are now bringing their own lightweight, flexible, and safe co-bots to market.

In 2016, co-bots represented five percent of the market for industrial robots, with an average price of around $30,000. Experts expect that the number of co-bots sold will increase by 60 percent from 2016 to 2021. That will bring major changes to small and medium-sized businesses. At the moment, only around 10 percent of these companies utilise robots, but within 10 years, that number is expected to increase to 60 or 70 percent.

Coping with variation is Amazon's challenge

Dealing with variation is the biggest trend in industrial robotics. A truly flexible robot arm should be able to deal with very different products, changing environments, and even reorganised production processes or new quality standards.

Take a warehouse of the e-commerce giant Amazon, for example. Amazon started as an online bookseller in 1994, then went on to sell videotapes and DVDs. Today you can order anything imaginable from Amazon: from teddy bears to T-shirts, and from toy robots to telephones. Amazon is also one of the biggest users of — and investors in — robot technology. Its warehouses have tens of thousands of mobile robots that transport stacks of crates filled with all sorts of products. They drive criss-cross through the warehouse along a network of paths that resemble the street grid of New York. The driving robots politely grant the right-of-way to those who need it, don't bump into one another, and efficiently bring their cargo from beginning to end of the warehouse's logistical process.

Eventually, the robots come to a human who manually removes the items from the crates, scans them, and puts them in a box. Yes, you read that right: the picking up and packaging is still done by hand. Naturally, Amazon would like robots to do that part of the warehouse work too, but they are still much too slow and they aren't good enough at recognising and picking up a random product.

That's why the company created the annual Amazon Picking Challenge in 2015 (renamed the Amazon Robotics Challenge in 2017). The contest's ultimate challenge is to remove random items from a shelf in a random warehouse and put them in a box faster than a human can. The robot doesn't know in advance where on the shelf the product is located and has to find it using image recognition. Then it has to think how it will grab the product from the shelf. Sometimes it has to move another product aside to reach the item located behind it. The robot arm is equipped with a gripper that consists of a vacuum suction cup that can hold a wide variety of products, providing they are not too heavy.

In 2016, a team from Delft University of Technology won the contest. Their winning robot arm took an average of half a minute to remove a product from the shelf and put it in the box. In Amazon warehouses, humans can do the same job in less than one second. Humans are also able to process hundreds of thousands of different types of products, while in 2016 the robot could only handle 50 pre-scanned items. Still, it is only a question of time before a robot will be able to grab and move random products faster and more reliably than the most skilled humans.

To do that, the underlying robot technology will have to make it possible for robots to deal with lots of variation. The most important technological trends driving robotics today are ever-increasing computing power, better and cheaper sensors, improved algorithms for computer vision and machine learning, robots that are increasingly interconnected, cloud computing, and lighter, better materials.

Combining these trends will make it possible for robots to become more mobile and effective in unstructured, unpredictable environments, such as homes, offices, and cities. Their potential applications are virtually limitless. The number of hospital transport robots is rising, along with warehouse robots, farming robots, lawnmower robots, vacuum cleaner robots, and robots for environments that are hazardous — or even inaccessible — to humans, such as disaster areas, the deep ocean, or Mars.

The science fiction author Arthur C. Clarke once said: 'When it comes to technology, most people overestimate the impact in the short term and underestimate it in the long term.' The same applies to robotics. In the short term, we often expect too much from robots, but we also tend to underestimate their massive impact over the long term. The household robot that does all of your chores is still a long way off, but the self-driving car has almost reached your driveway.

Like a ballet: robot car building

LIKE A BALLET: *Robotic arms move together gracefully in the car factory.*
VDL NedCar

The gigantic factory floor could be the setting of a science fiction film. Metres-high orange robot arms wave gracefully, like a choreographed ballet over dozens of steel car chassis sliding by on conveyor belts. In this plant, the robots move as if they were playing a match of Olympic synchronised welding. Sparks from the spot welds fly high into the air. Floor pans, wheel wells, doors, roof panels and bonnets — everything seems to flow from robot arms with millimetre and millisecond accuracy. The robots lift the 400 kilogramme car bodies as if were toys.

Under the leadership of our guide Jan van Daal, we and 30 other curious onlookers ride in a train through the factory operated by carmaker VDL Nedcar in Born in the Netherlands.

'The chassis plant is the most automated part of the whole company', Van Daal explains. '99 percent of the work here is done automatically, using around 1,000 robots. It's not true that robots only take people's jobs. They also create new ones. Plus, they do the work that people would rather not do. Welding used to be the hardest job at the factory.'

Since 2014, VDL Nedcar has built the MINI Hatch, the MINI Cabrio and the MINI Countryman for BMW. The latter two models are even made exclusively in Born. On 1 August 2017, the BMW X1 joined the production line. In total, 5,500 people from 33 nationalities work at VDL Nedcar today. It's only thanks to the far-reaching level of robotisation that the car factory can survive in the Netherlands. If there were no robots, automotive manufacturing would have become too expensive in the Netherlands, and it would have moved to low-wage countries in Eastern Europe or Asia years ago.

In the meantime, the train has moved on to the paint shop, where the bare steel chassis from the chassis line are cleaned, de-greased, and painted. Since the paint shop needs to be entirely dust-free, we aren't allowed to look around. The only bit of biology allowed inside are more than 10,000 emu feathers from Australia. Their anti-static characteristics make the feathers ideal for polishing the dust off of the cars.

A video shows how each of the cars are painted a different colour. Van Daal: 'Around 60 robots work in the paint shop. They know exactly what colour each body should be painted. Look at how they move. It all looks so nimble. I've programmed a lot of the paint robots, and each time you notice that it gets a little bit better. Yeah, programming is an art, just like painting.'

Finally, we ride through the assembly hall, the last hall in the production process. Painted chassis are moved from the paint shop to the mile-long assembly line. This is where most humans — and the fewest robots — work. Forklifts drive in and out. Metres-high shelves are filled with recently delivered parts. Here, Nedcar employees install almost 3,000 different parts to each car. Only the biggest and heaviest parts, like the front and rear seats, are installed by robots. Humans are still irreplaceable for the finer work, such as wiring, brake lines, airbags, and insulation material.

Every day, VDL Nedcar builds 700-800 cars. Van Daal: 'That number would be impossible without robots. We could turn out all the lights in the factory, and the robots would keep working at the same pace, because they don't have eyes.'

5 Learning to speak to people

The problem with machine talk

Standing on the table is a 60-centimetre humanoid robot, called a Nao robot. It's made of plastic, has a head with two eyes, a small mouth (that cannot move) and no nose. It has a torso, two arms, and two legs. Nao can talk and listen, dance, and play games, such as 'Guess the Sport'.

'I will play a sport', says Nao. 'Can you guess which sport it is?'

Nao imitates bouncing a ball up and down, points its eyes upwards, and pretends to throw a ball into the air.

'Basketball.'

'Yes, that is correct.'

It continues with the game. 'Can you guess which sport this is?'

Nao bends its knees and moves its hands, as if it were pushing itself with two sticks. The sound of crunching snow comes through its speakers.

'Skiing.'

Nao shakes its head. 'No, that is incorrect. This is not rowing, it is skiing.'

'But that's what I said!'

Nao doesn't understand that, however. It doesn't say: 'Sorry, I misunderstood you'; it simply ignores the protests and continues imitating the next sport.

Nao is a little like an idiot savant; a robot that can be programmed to speak dozens of languages and recite entire encyclopaedias from memory, but which has trouble communicating with people. But that's not necessarily a bad thing. Let the robot be a robot. Robots don't have to be identical copies of humans. Robots are allowed to make mistakes, and we should appreciate the things that they can do better than humans: they never lose their patience, they never get tired, they have a perfect memory, and they are never in a bad mood.

TEACHER WHO NEVER LOSES PATIENCE:
Nao teaches Dutch children English.
Paul Vogt

Nao is often used for education, such as helping children with diabetes deal with their disease by playing a game with them, where they learn about insulin and carbohydrates. Many children enjoy learning that way, and some are more honest with the robot than they are with adults. As a token of appreciation, some children even give Nao presents.

When it comes to human interaction, robots can already do quite a bit, but talking to a robot the way we talk to other humans is still difficult. First, the robot has to be able to recognise our speech. It has to know how to link sounds to syllables and words. Every person pronounces the same word slightly differently, but the robot needs to be able to recognise all those different sounds as representing the same word. Speech recognition is improving, but it is still far from perfect.

Secondly, the robot must understand what the words and sentences mean. To do that, it must know the grammatical structure of the language, and the meaning of words, phrases, and entire sentences. The robot also needs to possess knowledge of the world, because words refer to objects and concepts that have a meaning in real life.

It also needs to know how people use language in everyday life, because the exact meaning of spoken language is often different from the dictionary definition. When you tell your partner: 'the light in the living room was on', you probably mean: 'will you please turn off the light before you go to bed from now on'. And when you say: 'That's great' in a certain tone, you probably mean it sarcastically: 'I'm not happy about that!'.

Third, once the robot has understood what you've said, it has to think of what to say in reply, and how and when to speak. It has to generate speech the way humans do, with the right intonation, the right accent, and the right rhythm. A sentence shouldn't sound like a simple chain of words. We humans do that automatically, but what comes naturally to humans is often extremely difficult for robots.

Finally, the robot has to deal with the fact that language is constantly changing. New words are created — 'googling' — and older words disappear — 'cottier' used to mean a rural labourer living in a cottage, and 'coxcomb' is a vain or conceited man. Existing words can gain new meanings — 'cool' — and new grammatical constructions and sayings are born — 'the future ain't what it used to be' or 'you wouldn't have won if we'd beaten you'.

SHRDLU! The first experiment in robot conversation

The first ground-breaking experiment to study how we can talk to robots was conducted by Terry Winograd. From 1968 to

1970, he conducted his PhD research at Massachusetts Institute of Technology, where he developed a computer program that would allow the user to communicate with a robot using natural language. His program was called SHRDLU, a nonsense word Winograd made up. SHRDLU is a virtual world on a computer screen, consisting of a virtual robot arm and simple geometric objects on a table: coloured blocks, pyramids, and balls of various shapes and sizes, along with a box.

The user can give the program commands in English, such as 'pick up the large red block'. The virtual robot arm will then perform the command on the computer screen.

The user can also ask questions about the blocks. 'How many blocks are in the box?' And the robot will give the correct answer.

In SHRDLU, the user can also ask a question that requires basic physical knowledge: 'Can a pyramid sit on top of a pyramid?'

Initially, the robot answers: 'I don't know.'

Then the user can tell the robot to try it: 'Stack one pyramid on top of another.' The robot arm tries to perform the command, but naturally fails. In so doing, however, the robot learns something new, and subsequently answers that one pyramid cannot sit on top of another.

When it came out, SHRDLU was seen as a major breakthrough in the field of artificial intelligence. Winograd's experiment showed that it was possible for a robot to develop an understanding of a simple world, and to communicate about that world using natural language. But Winograd quickly concluded that the SHRDLU approach was a dead-end. SHRDLU works because it involves a miniature world with a limited number of simple objects, and a limited number of commands that the robot arm can perform. But that approach proved impossible to scale up to the real world, with its endless number of possible objects and actions, which are often difficult to reduce to simple rules.

Toilets are hidden: translation problems

Paul Vogt, Associate Professor at the Centre for Cognition and Communication at Tilburg University in the Netherlands, works on communication between humans and robots. He studies how people and machines can anchor the meaning of language in the world around them. He also studies the ways that humans and robots can learn a language: humans learning from humans, robots from robots, but also humans from robots and robots from humans.

Vogt: 'In the SHRDLU experiment, the communication was programmed in advance, but the real world is far too big for that. Robots will only understand human language once they can learn the language for themselves. We humans spend our entire lives learning language bit-by-bit. We learn something with every interaction we have with the world around us. The learning models we've developed for robots only work well when we have a large amount of data. A robot can use a lot of pictures of a dog to learn how to recognise a dog, but in the real world, a robot will frequently come across situations for the first time. So you need learning models that use very little data, and we don't have those yet.'

But what about virtual personal assistants, like Siri, Apple's smartphone assistant, or Amazon's Alexa, the intelligent home speaker? Can't we talk with them at a basic level when we want to know the weather forecast or what is on our agenda for the day?

Vogt: 'Computers will never have a real understanding of the world, because they can never actually experience the real world. At most, they can gain an understanding of a virtual world. They can talk about coffee, but they don't drink coffee, and they don't have a sense of taste.'

Vogt's PhD supervisor, the Belgian artificial intelligence

pioneer Luc Steels, had this to say about the problem: 'The main limitation to artificial intelligence today is meaning. All of the applications that exist are based on avoiding meaning. Take automated translation, like Google Translate, for example. The program doesn't try to understand what's being said and to express that understanding in a different language. No, there are databanks of text pairs, found on the Internet. One piece of text is linked to a comparable piece in another language. That happens on a huge scale, and sometimes it produces bizarre results.'

That's how a Dutch restaurant owner came to hang a sign on the toilet door saying: TOILETTEN ZIJN VERSTOPT (TOILETS ARE CLOGGED), with Google Translate's English translation underneath: TOILETS ARE HIDDEN.

Robots have an advantage over computers, because they have a body and sensors, they can move around in our world, and they can interact with it. In principle, they should be able to learn to understand our language as well.

Vogt agrees: 'I definitely see a future in communicating with robots using ordinary human language. The holy grail would be for robots to be able to learn to understand a human language themselves, and for us humans to be able to communicate with them in a mature, natural way. But it will be a long time before we can talk to robots the way we do with other humans. At the moment, the future of talking with robots is painted in far too rosy colours.'

People develop language comprehension by seeing relationships between the world around us, Vogt explains. 'People do that by doing things. I have a porcelain mug of coffee. You have a plastic cup of coffee. They hold the same thing, but they are very different objects. Yet we can differentiate between these different objects and other objects, like a table or a chair. We have learned that we can drink coffee from them, but we also know that you can do totally different things with them. When you're angry, you can throw the cup of coffee at the wall. We also know that we can use the cup in new ways. You can hold paperclips in it, for example. People have learned all of those things through their interactions with the physical world.'

So what is it that makes it so difficult for robots, exactly? According to Vogt, it lies in a combination of all of the skills a robot needs: its powers of observation, its capacity for thinking, and its ability to act. The sensors aren't yet good enough, the learning models aren't yet good enough, and the mechanics of the robot body are nowhere near good enough to be able to deal with its surroundings as easily as humans. And we haven't even dealt with feelings and emotions yet, although we'll do that in the next chapter.

As long as a robot can't deal with the world in a natural way in a physical sense, it will be difficult for it to make a link between a word and its meaning in the world as easily as humans do. Luc Steels describes it this way: 'Take the concept of a door. When you read the definition — an opening in a space that can be closed, with a latch etc — you can form an idea of a door, even if you've never seen one before. A robot can't do that. You can show it 20 examples, but when it's confronted with a door that's different, it won't recognise it as a door. We also link actions to concepts: opening a door. If you can't do it the normal way, for example because it doesn't have a doorknob, then we'll find another solution. Robots can't even come close to doing that.'

A robotic teacher never runs out of patience

Vogt sees many benefits to natural communication between humans and robots. In the European research project L2TOR (pronounced like the Spanish words 'el tutor'), he studies whether robots can help five-year-old children to learn a second language. Like many other educational projects, Vogt uses the Nao robot for this purpose.

In one of his experiments, Nao tries to teach Dutch children the English names for animals by saying things like: 'I spy with

CHATTING WITH A ROBOT: *Author Bennie Mols with Pepper.*
Irma de Hoon

my little eye, and it's a... monkey.' At the same time, it performs movements that resemble those of a monkey. A tablet then shows pictures of four different animals, including a monkey, and the child has to touch the correct picture. If the child picks the right answer, Nao says: 'Well done.'

Vogt: 'We conducted this experiment with 80 children, and we showed that it was a fun way for them to learn new English words. It even seems like children learn better from a robot than from a book or tablet. The hypothesis is that it's because people have evolved to deal with other people, and that the physical presence of another person is important in order to learn a language. Even if that other person is a robot.'

Vogt adds that human teachers are still better than robot teachers. 'The robot isn't there to replace the human, but as an extra tool for when the human doesn't have time or has to teach too many children at once. In my vision for the future, a teacher might say to a child for whom classical lessons are too hard, or even too easy: 'Why don't you practice with the robot for half an hour.' Then the robot would recognise the child, know what level the child is at, and what lesson he or she had last time. The robot never gets tired, never loses its patience, remembers exactly how far each child has progressed, and adjusts its lessons precisely to the individual.'

Vogt encountered some obstacles in this simple experiment, however. For example, the robot's speech recognition didn't work for younger children. 'I thought it had improved, but that was disappointing. Speech recognition works fairly well for adults, but it has trouble with young children because they often don't speak grammatically and use unexpected words, and they also pronounce words wrong. In order to compensate for that, speech recognition software needs a lot of data about child speech, and we don't have that data available yet.'

Since the robot's speech recognition still isn't up to the challenge, the children have to touch a picture of a monkey on a tablet when they hear the English word 'monkey', instead of saying the word 'monkey' back to the robot.

A second problem is the way the robot pronounces words. 'It's very difficult for the robot to put the right accent on syllables or words. Another factor that often sounds unnatural, is the time gap between two consecutive words. A small gap makes a big difference, and children notice immediately when a robot doesn't talk naturally. We're working with the developers of the robots' software in order to make them talk more naturally, but that's much more difficult than many people think.'

A third problem deals with the robot's predictability. Vogt: 'In the beginning, the children thought it was fun and exciting

DEVELOPING A ROBOT LANGUAGE: *Like humans, robots can categorise the outside world and make up their own words.*
Luc Steels

to talk with the robot. But after they heard it say the same thing over and over — 'I spy with my little eye, and it's a...' — we noticed that they started to get bored. The robot may have infinite patience, but the children don't. At a certain point, the enchantment wears off.'

Robots therefore need to be able to improvise a bit in order to ensure that we humans can build a social-emotional bond with them. Or, as robot researcher Guy Hoffman, who studies the interaction between humans and robots, puts it: 'Robots need to resemble chess players less, and actors or musicians more. Robots that are less than perfect are perfect for us.'

Robots have learnt to communicate with other robots

We humans want to teach robots a language because we want to be able to talk to them as easily as we do with our fellow humans. That way, we can give them instructions in everyday language, instead of having to press buttons. But could robots develop their own language, so that they can talk about what's important to them based on how they see the world? Vogt and Steels have worked together to study that question.

Vogt: 'We mainly did that to study how language could have developed over the course of humans' evolutionary history. For example, we put robots equipped with light sensors into a box with four sources of light. Then they drove around the box around according to basic, pre-programmed strategies. The robots also exchanged information about the world they could perceive. After a while, they developed words for each source of light: 'huma' for one light source and 'kyga' for another. This experiment showed that robots start to categorise the outside world, just like humans, despite the fact that each robot developed a slightly different representation of a light source in its robot brain.'

Practical applications of this type of research are still a long way off, but they do appeal to the imagination. Vogt: 'If we equip robots with sensors that we humans don't have, like an infrared sensor or an ultrasound sensor, then they can observe a world that humans haven't developed words to describe. That could be useful in environments where people can't go, or only with a great deal of effort, such as Mars or the bottom of the ocean. You could imagine that in such extreme circumstances, robots might develop their own language and communicate with one another in that language.' Vogt laughs and adds: 'But robots might think it's easier to communicate directly from brain to brain, instead of doing it through language.' That way, one robot can send what its brain thinks and feels, as expressed in noughts and ones, directly to another robot brain without having to convert those noughts and ones into a spoken language first.

Whether robots develop their own language or only use language to talk to people, language is about more than just communication. Language is the foundation of cultural evolution. By using language to explain something to another person, the other person can learn new things faster than if they had to find out everything for themselves. Language stimulates innovation and creativity, and makes it possible to communicate about abstract concepts. When we give our thoughts free rein, we can imagine a future in which robots communicate with one another in their own language, develop their own culture, and ask themselves why humans understand so little of what they say: 'Klaatu barada nikto.' (Google it!)

Making a stubborn, annoying, talkative robot

You might know animator and video artist Jo Luijten from his viral YouTube videos about how Angry Birds and Facebook would have looked in the 1980s, but he spends most of his free time building a robot. As a linguist, Luijten is most interested in the speech interface of his 'Jobot'.

'I see my robot project as a hobby that got completely out of hand. The only goal is to see if I can actually make it work.' That goal is not to create a complex robot. 'My robot has to make people laugh. It has to be an insufferable, pig-headed, irritating, chatterbox robot that monopolises the conversation. The neighbour you don't want to come across in the supermarket.'

ANNOYING AND TALKATIVE:
Jo Luijten wanted to make his Jobot behave like an annoying neighbour.
Jo Luijten

Luijten taught himself how to program in BASIC in the 1980s. Since then, he has worked on all sorts of creative computer programs, often with sounds and music, but in the end he decided not to study Computer Science. 'I rarely actually completed a program. Maybe because I had too many ideas, and I had already started the next project before the last one was finished. The fact that I didn't feel like it anymore once the programming became too hard probably also played a role.'

Today, he spends much of his time programming and tinkering on his robot. 'I have more patience now than when I was younger. It's also easier now, because you can use functions that are already part of the operating system. Now, your program can call on Windows' built-in speech recognition and speech generator functions.'

Luijten programs his robot in Visual BASIC, with a text file that indicates how the robot should react to specific input. 'I try to make the program seem smart by using a huge database of pre-digested sentence fragments and replies. For my experiment, that's more important than real artificial intelligence. Like the irritating neighbour, my robot is someone who doesn't listen well, and that might make it seem more real. When the robot talks a lot, it seems smarter. I also want my robot to bring up topics for conversation on its own. I've noticed that most robots, both hobby robots and professional robots, never start a conversation. That's something people do. My robot chooses a random number, which is linked to a subject in the database, so the robot might suddenly start talking about something completely different. When it comes to that, the robot is a lot like me.'

Luijten uses clever tricks to make the robot seem more communicative, such as 'cold reading'. 'I wanted to include techniques like those clairvoyants use in my program. Sometimes they can learn quite a bit by asking very clever questions. For example, if someone is in their thirties, there is a good chance that they don't know what they want to do with their life. They've got their degree, live on their own, and they think: is this it? In that case, the robot can say: "I have the feeling that sometimes you don't know what to do with your life. You've got everything: a job, your own apartment, and you have a relationship. Am I right? Don't worry. You'll figure it out eventually." If the robot is right, it seems like the machine has social skills.'

Luijten also built the robot's hardware himself. 'The robot's body is an aluminium housing, with a Plexiglas dome for a head. It looks a bit like R2-D2.' The robot is equipped with a camera, speaker, and microphone. Two lamps are installed under the Plexiglas dome. Luijten: 'I installed them because it seemed like the robot didn't have eyes. The stereo camera is hard to see under the dome. The lamps have two functions: now people know where to look at the robot, which is handy for facial recognition. And when the lights are on, it indicates that the camera is on as well.' Other lights glow when the robot perceives sounds or when it talks. 'Last year, I downloaded software that lets me digitise my own voice, so Jobot talks in my own voice. But I haven't had time to work on the project recently, because now I have a real flesh-and-blood baby to take care of.'

Will we all have a robot at home within a few years? 'That depends on how you define a 'robot'. I think that the smart refrigerator, the smart coffee machine, and the self-driving car and things like that will become very common. Just this morning, my coffee machine reminded me that I had to clean one of the parts. When I was young, that would only have been possible in a science fiction film.'

6 Robots get emotional

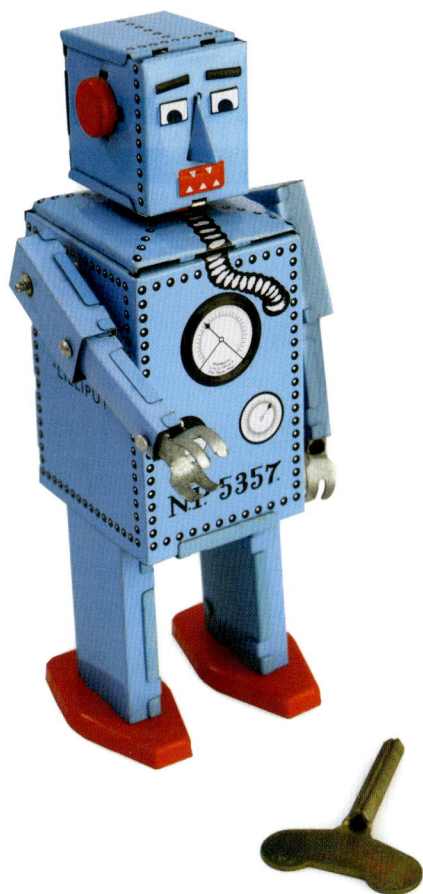

Emotional robots encourage humans to interact with them

We're all familiar with robots that can display emotions from books and films. C-3PO, the neurotic whinger from *Star Wars*, is more emotional than his human colleagues. Samantha, the operating system from the film *Her*, says that she is in love with the protagonist Theodore. And perhaps the best example of an emotional robot Marvin, from *The Hitchhiker's Guide to the Galaxy* is chronically depressed. Emotional robots are often humorous, but they can also be very creepy: *The Hitchhiker's Guide* has cheerful sliding doors which enjoy serving people, opening and closing with a self-satisfied sigh.

An emotional robot? Help! Hopefully, we won't have to deal with moaning doors at the supermarket anytime soon. And adding emotions to real robots quickly conjures up negative images: arguments with your robot vacuum cleaner, factory robots organising a strike because they don't feel their tireless efforts are appreciated, or a self-driving car that is afraid to go on the motorway after an accident.

No, not every robot needs emotions. They have little functional value for a simple vacuum cleaner or automatic doors. Yet emotions may be useful for some robots: for example, a robot that is able to recognise your emotions and adapt to your needs, which might not always be rational. How wonderful would it be if your care robot could notice that you're feeling grumpy, and bring you a cup of tea to cheer you up?

Robots that display emotions look more real and lively, seem more convincing, and make people feel more attached to the robot. Robots that can mirror the user's emotional state — by appearing happy or sad when the user is happy or sad — receive

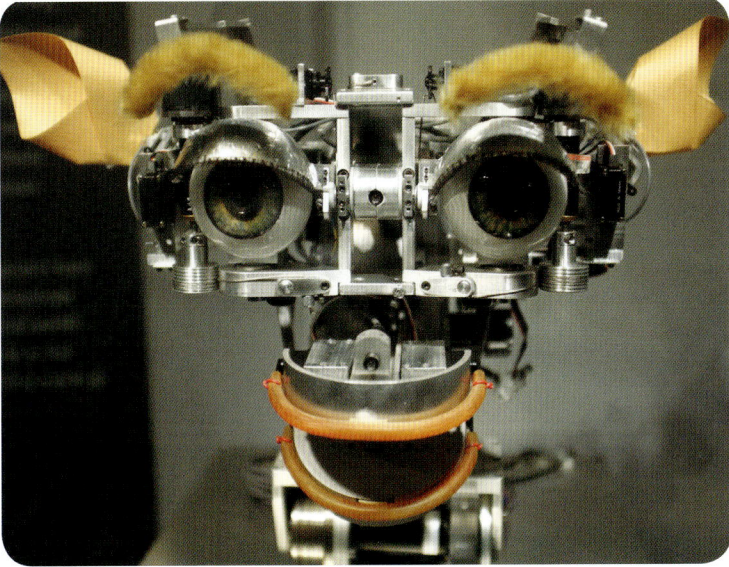

FACIAL EXPRESSION: *Cynthia Breazeal designed Kismet, the first robot head to show emotions.*
Bennie Mols, MIT Museum

more positive reactions. It would also be useful for a teaching robot to be able to recognise emotions, so that it can adjust its teaching style when it notices that a student is bored, interested, or frustrated.

Cynthia Breazeal is a pioneer in the field of social robots at Massachusetts Institute of Technology. Since the 1990s, she has conducted research into the interaction between people and robots. In 2000, she built the robot Kismet, the first robotic head that can display emotions. Breazeal: 'I study how artificial intelligence can contribute to making humanity flourish. To do that, you need emotions and human relationships. That doesn't mean replacing people, however. Why would you? We are social beings, we want to belong to a human community, feel appreciated by people, and feel that we are needed. It does matter that you are a human, and not a robot. But that doesn't mean you can't also have a robot that adds value to your life.'

When Breazeal was doing her PhD research at MIT in the 1990s, researchers rarely thought about social or emotional robots. 'At the time, research into autonomous robots only dealt with things that were far away from people's daily lives. The moment that I started seriously thinking about it was when NASA landed the Sojourner rover on Mars in 1997. We send robots into dangerous situations, even all the way to Mars, but we still don't have them at home. Why is that?'

A lot has changed over the past 20 years. In 2014, Breazeal presented a prototype of her new companion robot Jibo, which will soon be available commercially. Jibo looks like a kind of chubby desk lamp, with a round screen as a face and a hemispherical head. It features a camera equipped with facial recognition, and it understands spoken commands. Breazeal: 'Today, we understand much more about social robotics. We understand when and why we feel personally involved in an interaction, but we also have much better interaction models for education, health care, care for the elderly... We understand more about the psychology of humans interacting with a social robot, and how much it differs from interacting with a screen. Imagine that you build a robot that teaches a child to read and write. Research has shown that the child learns better when the robot displays the right emotional and social behaviour.'

Breazeal's team has invested considerable effort into creating Jibo's fluid movements, which represent familiar 'body language'.

'The body is very important, from a social perspective. Our brains have evolved to communicate physically with others, and our mind has adapted to that. Interaction with a screen just doesn't have the same depth to it.'

A robot can work out how you are feeling

At some point during our evolution, humans developed emotions. We don't know exactly how and when, but they probably evolved along with cognition. In short, cognition is responsible for observing and organising the world around us, while emotion helps us to evaluate it — is this situation positive or negative? — and to make decisions. Emotion produces a kind of basic reaction, which allows us to react faster and easier: running away because you are afraid of a predator, or a desire to eat, so that you start looking for something nutritious in time.

Even though we might think that thinking and feeling are opposing concepts, they aren't really. They are very closely related, and are constantly influencing one another. People who can't feel emotions as strongly as they did before a stroke, for example, also appear to make poorer decisions, and to take longer to make them. Emotions help us to make choices, which can also come in handy for a robot. Even aside from the finer points of interaction with humans, emotions could help robots to function better in complex, unpredictable environments.

Humans have lived with emotions for so long that we are very good at interpreting facial expressions. When Beethoven had been deaf for a long time, he claimed that he could see from the musicians' facial expressions whether they were interpreting his music correctly. Of course, emotions can differ from person to person and from culture to culture. Some people have a face that always looks angry — sometimes called the 'resting bitch face' — and in Southeast Asia, it is fairly normal to smile when you feel sad or ashamed.

Nevertheless, researchers report positive results for computers recognising emotions: computers are fairly good at determining what emotion a person is feeling from their speech, facial recognition, and physical characteristics such as heart rate. They're not perfect, but they are almost as good as humans at recognising emotions. In order to recognise emotions from photographs of faces, computers use the same kinds of techniques used in computer vision. Extremely fast cameras even make it possible to recognise subtle micro-expressions: minor facial expressions that only last a fraction of a second, and are

UNDERSTANDING HUMAN PSYCHOLOGY: *Jibo is a social robot that can see, hear, speak and learn* Jibo, Inc

often a sign of unconscious or suppressed emotions.

Being able to recognise emotions is useful in many areas other than robotics. Take security, for example: cameras that can detect aggression. But there are also commercial applications: perhaps your future television can recognise your irritation or boredom, and learn what kind of programmes you like — or don't like. And companies love to be able to analyse whether people are talking positively or negatively about them on social media.

Why am I afraid? Understanding human emotions

To create robots able to deal with feelings, it would be helpful to have a clear definition of different human emotions — and that presents researchers with a challenge. Psychologists don't entirely agree which basic emotions we have, because the emotional spectrum is complex and specific to individual people and cultures. The most frequently listed basic emotions are anger, fear, disgust, surprise, joy, and sadness. But makers of emotional robots often look to a wider model of emotions. One commonly used model differentiates no fewer than 22 emotions, each accompanied by a handy list of specifications.

For example, the model specifies 'fear' as '(displeasure about) the prospect of an undesirable event'. The strength of the emotion 'fear' depends on two characteristics: the degree to which the event is undesirable, and the likelihood of the event. In theory, your vacuum cleaner robot could be afraid of falling down the stairs — which would be an undesirable event for it — but it would not experience that fear in practice, because its sensors are so good they make such a fall highly unlikely. These are relatively simple rules, and are therefore useful to program into a robot brain.

So does a robot really feel emotion? Or does it only conclude that it should have an emotion, based on its situation and a number of rules? That is a more difficult question, which deals with issues of consciousness, physicality, and humanity. What does it mean to have or feel emotions, exactly? To what extent does one need a biological body to have emotions? Or can a robot body also have emotions?

We humans have a real talent for ascribing personalities and emotions to objects: a computer 'has a bad day', the coffee machine at work has 'its own will', and this plant needs some kind words now and then to make it bloom as enthusiastically as the others on the window sill. The same applies to robots as well: we are quick to believe that they have emotions.

Help! My robot looks angry

Researchers like David Hanson try to build human-like robot faces that display emotions as realistically as possible. That's not easy, because humans have dozens of muscles in their faces, and it's difficult to imitate facial expressions accurately, as anyone who has seen a poorly-acted film can tell you. Hanson only simulates the most important muscles in his robots, which can mean that facial expressions aren't quite realistic.

To avoid the challenges of imitating facial expressions, many researchers are using body language to show emotion instead: a hanging head for sorrow, raised hands and hunched shoulders for surprise or fear. Many humanoid robots, like Nao and ASIMO, express emotion mainly through body language, and not facial expressions. And it seems to work: we understand robots better and feel they are more convincing if they show something about their emotional state through body language — regardless of whether they actually have any emotions.

Cynthia Breazeal believes that robots don't have to look like people in order to be friendly and engaging, and that even their expressions of emotion don't necessarily have to look like ours. 'Facial expressions don't have to be realistic. We can read expressions and emotions in all sorts of things. People are exceptionally good at that. That's why we can look at a dog and understand what it's feeling.' Being less rigid in how robots display emotion may also make it easier to escape from the uncanny valley.

Breazeal also finds inspiration in the rich history of animation. 'None of my robots look like people. We don't like characters that are almost-but-not-quite human. We build relationships with dogs and cats just like we do with other people. We all grow up with stories and cartoons with non-human characters. That relationship is special.' Another thing we see in stories is that robots don't have to be perfect. 'The most interesting characters are 'perfectly flawed', which is why we empathise with them.'

Let us return to Kismet, the robotic face Breazeal built in 2000. Kismet consists of a neck and head on a metal box. The robotic face has large eyes, hairy eyebrows, pig-like ears, and a mouth. Kismet's facial expression is determined by the 'motivation system' that Breazeal developed, in which visual input from Kismet's stereoscopic camera is converted into the emotions it displays.

Kismet has the emotional development of an infant. Breazeal: 'That seemed to me to be a reasonable starting point. As babies, we don't yet have a fully developed brain or emotions.' Kismet reacts to what happens around it in a very basic way. You can make it happy by playing with it: holding or moving blocks in front of its face. But if you move the blocks too fast, or bring them too close to its face, it becomes afraid or angry. On the other hand, if there is not enough action, it gets bored and its eyebrows and ears begin to droop.

Kismet is made for interaction with humans, but it looks nothing like one. That wasn't Breazeal's intention, because she developed Kismet mainly to study how the interaction between humans and robots works. 'When I started developing Kismet, it was the very first social robot in the world; the first robot that was truly designed for face-to-face interaction with people. And that raised questions like: how can you give a machine social intelligence?'

A robot's facial expressions and body language therefore don't have to look like those of a human in order for us to understand them. You could think in much simpler terms still. For example, by putting a tail on a vacuum cleaner robot, like researchers at the University of Manitoba have done. A vacuum cleaner robot doesn't evoke emotions, of course, but since people are so used to seeing emotions in pets, they can be a way to show the status of the robot in a very simple and intuitive way. Test subjects can interpret a wagging or hanging tail on a vacuum cleaner robot without difficulty. It doesn't even matter whether or not they have pets themselves.

Establishing a bond with your robot

Cynthia Breazeal considers building a long-term, close relationship with a robot to be a major challenge. 'The intense, human interaction with a robot is crucial. That would be the 'killer app' for social robots.' Achieving that is one of the goals that Breazeal has set for her new robot Jibo, but she acknowledges that there is still much to do. 'Jibo is the first of its kind, and the technology still needs to develop further. How do you build a relationship with an artificially intelligent being? That will remain a challenge for the foreseeable future. To do that, we have to thoroughly understand how people understand and interpret one another's behaviour, and how we derive someone's mental state from their actions. Only then can we make machines that really understand us.'

But on our way to reach that goal — fully comprehending relationships between people — there are already all sorts of other relationships that we do understand better, explains Breazeal. 'We humans have all sorts of different relationships. The relationship between humans and dogs, for example; we now know how that works. I also understand the relationship that I have with devices and technology; a kind of utility relationship. But we still don't completely understand the relationship between humans and robots. It won't be the same as our relationship with another human. There will also be many differences between individual people: some are more hesitant in their dealings with robots, while others go all-in. And we will have to make sure that the robot will be able to deal with that.'

Breazeal once met a child who brought a gift for a robot. Is that strange? 'Not at all, there's nothing wrong with that. The child knows full well that it's a robot, but she was practicing her empathy. In that way, technology can help us develop our capacity for being more humane to one another. Based on the research I've done, I believe that it's possible; that technology can also help us to find bonds to other people. I call it 'warm technology'. A robot like that isn't just user-friendly; it's actually friendly.'

HitchBOT: a globetrotting social media star

Hitchhiking is a dangerous way to travel, but in 2014 and 2015, Canadian researchers Frauke Zeller and David Harris Smith sent their robot hitchBOT out to hitchhike around the world on its own.

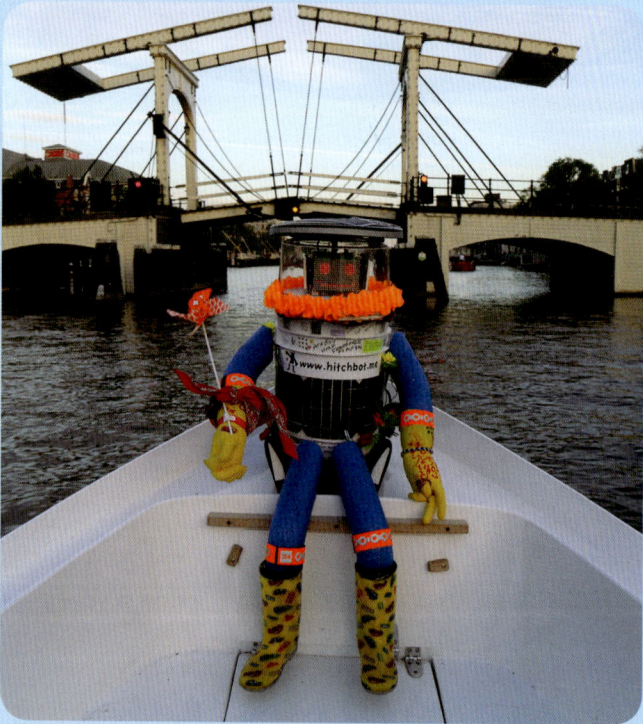

FREE-SPIRITED TRAVELLER:
HitchBot, pictured in Amsterdam,
quickly became a social media star.
David Harris Smith

Zeller and Harris Smith were interested in 'cultural robotics': studying how people react to robots in places where they least expect them. Zeller: 'It teaches us something about how humans see robots and artificial intelligence in society. We wanted to use that approach to think about the concept of safety. So we came up with a hitchhiking robot, which could have conversations with the people who picked it up. You often hear people ask whether we can trust robots, but we wanted to turn that question around: can robots trust people?'

Zeller explains that the robot had to look cute and trustworthy. 'You won't pick up a robot if you don't trust it. The robot had to be likeable, and look like it needed help. That's why we made it around the size of a child. It even had its own built-in car seat so you could secure it in the car.' Drivers could pick up hitchBOT, carry it along for a while, then drop it off on the side of the road.

Other than that, hitchBOT couldn't do very much, except tweet its location and photos and conduct simple conversations. 'It mainly had to be friendly and fun. At first, we thought about making it a nerd, like the character Sheldon from the TV comedy *The Big Bang Theory*. But in the end, we didn't have the budget to install an advanced artificial intelligence, and we realised that people might not want to sit in a car with Sheldon for more than an hour.'

HitchBOT quickly became popular on social media. 'We started tweeting from the hitchBOT account even before the robot was finished, and at a certain point I noticed that we already had 100 followers. I thought that was really great! Later on, an editor at *The Atlantic* found out about us and wrote an article, and then it really took off.' Eventually, hitchBOT had almost 55,000 followers on Twitter. 'The traditional media quickly picked it up, which was lucky for us: it was a slow news period, and we had a feelgood topic. We really hadn't expected it to be that popular."

HitchBOT travelled thousands of kilometres through Europe and Canada. It even received a lift on a boat through Amsterdam's canals. But when it got to the US, its journey came to an end: in August 2015, the robot was found broken beyond repair in an alley in Philadelphia. Zeller: 'At first, we were shocked and saddened. When we set up the project, we naturally thought about what we would do if hitchBOT was destroyed. It was just part of the project. But the project was so successful, and we had earned so much attention around the world, that we suddenly had to go into crisis communication mode. In the end, we drew up a kind of obituary. After we got over the initial shock, we reminded ourselves that this was a natural part of the project. But to be honest: we really missed him. We would never receive tweets or photos from him again. The public's reaction really helped us. There were so many people who sympathised, who let us know that they had to cry when they read the news. The Mayor of Philadelphia even contacted us. People got inspired too: they sent us comic strips, songs, photos, and hitchBOT costumes.'

So what was the final conclusion? Can robots trust humans? Zeller doesn't doubt for a second: 'Absolutely. We don't know exactly what happened in Philadelphia, but that's not the important thing. The most important thing is the thousands of people who supported hitchBOT. During his travels around the world, hitchBOT became a symbol for hope, cooperation, and trust. The project showed that technology helps bring us closer together. You can compare it to the telephone: at first, people didn't want it in their homes, but eventually it became an important way for us to connect with one another. Maybe robots will do the same.'

7 Humans need robots and robots need humans

Meet the robot psychologists

In his anthology of short stories, *I, Robot,* science fiction author Isaac Asimov gave a major role to robot psychologist Susan Calvin, 'the first great practitioner of a new science'. When a robot behaves differently than expected, the robot psychologist is called in to help. Asimov's robots are made to serve humans, but they have their own personalities and regularly find unexpected loopholes in the ethical laws of robotics that they are programmed to obey. In the stories, Susan Calvin is a humourless nerd who has more in common with robots than with other humans: 'I've been called a robot myself.' When asked if robots are so different from humans, she answers: 'Worlds different. Robots are essentially decent.' The fictional robot psychologist has her work cut out for her understanding the complex interactions between humans and robots.

A robot psychologist may sound like a futuristic profession, but several people do a similar job today, such as the Americans Leila Takayama and Amber Case. With her background as a cognitive psychologist, Takayama has studied the interaction between humans and robots. She runs Hoku Labs in California, where she studies how people deal with their robots in practice on behalf of robot manufacturers. The journal MIT Technology Review selected her for inclusion in its list of the 35 top innovators under 35. Takayama is also listed as one of the '25 women in robotics you need to know about'.

Amber Case calls herself a cyborg anthropologist. She maintains that our intensive, everyday use of smartphones, tablets, computers and robots has made us all cyborgs now. Case works at the renowned MIT Media Lab and the Center for Future Civic Media, where she studies how people deal with digital technology, including robots. She is a frequent guest speaker at technology conferences, and is the author of two books: *An Illustrated Diction-*

ary of Cyborg Anthropology and *Designing Calm Technology*.

We spoke to Amber Case and Leila Takayama to find out: how can humans and robots best work together and what does that mean for the design of the robot?

Under-promise and over-deliver performance

Leila Takayama seems to be the opposite of Susan Calvin. She is open, friendly, laughs a lot, and has a great sense of humour. But she doesn't like science fiction. 'I thought science fiction was too masculine. It's often so stereotypical about men and women. I lost interest at a young age.'

At a job fair in California, Takayama ran into the people from the robot firm WillowGarage, the now-defunct makers of the robot PR2. As they talked, Takayama learned that the company needed a psychologist to study the interactions between users and PR2.

She fell in love with robots when she was asked to work on a telepresence robot, 'because that kind of robot is really about the interaction between two people via a robot. So the robot has to be as invisible as possible.'

A telepresence robot is basically a tablet screen connected to the Internet and mounted on a stick. It rides around on two wheels, like a kind of Segway. Say that you can't attend a meeting in person; with the remote-controlled robot you can still participate in the meeting from another location. Your face is shown on the computer screen, and you can see what's going on and contribute to the discussion, and even move around among the other guests during the reception afterwards.

Takayama: 'When you design a telepresence robot, you're faced with questions like: 'how tall should it be?' and 'what should it look like?' Those are the kinds of questions I studied.' Around the same time, *The Big Bang Theory* approached her because the producers

LIKE YOU ARE IN THE ROOM: A tele-presentor robot allows you to communicate remotely and even 'attend' post-meeting drinks.

Flickr.com

wanted to use a telepresence robot in the TV series. The result was the now-famous 'Shelbot'.

Since then, Takayama has spent almost a decade studying the interactions between humans and robots, and she has drawn a firm conclusion: 'Robotics has a tradition of promising too much and delivering too little. We were promised a robot that could do take care of all of our household work, but the best we have at the moment is a robot vacuum cleaner.'

According to Takayama, robot developers should be more honest in saying what their robots can do well, and what they can't. For example, she has studied how people first use the robotic toy dinosaur Pleo. Pleo was designed to imitate the behaviour of a week-old baby dinosaur. It can listen, make noises, and walk.

In her experiment, some users were told: 'Pleo is just as temperamental in what it thinks and how it feels as you are.' Others had their expectations lowered by being told: 'This robot has a limited ability to understand you and to interact with you.' The users were then given an opportunity to play with the robot. They talked to the toy dinosaur and watched how it reacted.

Takayama found that the users with the lower expectations were happier about their experiences than the users who started with high expectations. She applies this conclusion to robots in general. 'It's crucial to set realistic expectations about robots, and absolutely not to promise too much. The best thing you can do is to promise less and to deliver more.'

Takayama believes that a robot that looks like a human will set expectations that the robot cannot possibly meet. 'The fact that a robot is geared towards humans doesn't mean that the robot should also look like a human', she says.

Aside from setting more realistic expectations about robots, she also believes that robots should become better at communicating what is going on in their heads. 'Say a robot moves towards a door, and then stands still. People nearby don't know what the robot wants. Is it thinking about how to open the door, or does it want something completely different? The robot is moving around in our space, so it should adapt to us, and not the other way around. It would be easy to imagine it displaying a kind of thinking-emoji-face on a screen as it looks at the door. That would make it clear to people that it is thinking about how it should open the door.'

Silicon Valley utopias vs calm technology

In contrast to Leila Takayama, cyborg anthropologist Amber Case is well-versed in science fiction. In fact, Case believes that to understand how humans deal with robots today, we cannot ignore the collective mythology about robots created by science fiction. 'Apparently, in the early 20th century, people had a strong need to personify the mechanisation that came with the industrial revolution. That personification came in the form of the robot that walks like a human, talks like a human, and behaves like a human. The humanoid robot is a legacy of science fiction books and films.'

Case says that the legacy of science fiction has led us to expect too much from robots. 'The fact that we have a biological body that is born, grows, changes, then dies, will always differentiate us from robots', she says resolutely. 'Always.'

Like Takayama, Case also thinks that robot developers should set more realistic expectations. 'They need to say: these are our robots' limitations. Then they should answer the question of how users can get the most out of the robot within those limitations. And they shouldn't build robots for tasks that humans can do better. A robot waitress or waiter? I don't think that would be a success. The social aspects of the work, such as having a chat or showing a smile, are things that humans do much better.'

Case is an adherent to the concept of 'calm technology'. By that, she means technology that does not constantly push itself on people, like many smartphone apps which are designed to hold our attention for as long as possible. Calm technology stands on the sidelines as much as possible, and allows humans to lead the lives they want to live. 'In the 21st century, attention is the most precious commodity. That's why calm technology shouldn't demand all of our attention, but only a little, and only when it's really necessary.'

Her favourite example is the kettle. 'You put water in the kettle, leave to go do something else, and once the water boils it lets you know. You don't have to wait next to the kettle the entire time.'

When it comes to robots, Case feels that the robot vacuum cleaner and robot seal Paro are good examples of calm technology. Paro is used as a therapeutic care robot for elderly people suffering from dementia. 'Both of them give a little feedback to the user: not too much, not too little, but just right. The robot vacuum cleaner makes a cheerful sound when it's done with its work, and a sad sound when it gets stuck. That makes it cute. The robot seal doesn't pretend to take care of the elderly person; it makes the person feel better by letting them take care of it instead. When you pet Paro, it makes noises, moves its head, and blinks its eyes.'

An important principle in the design of calm technology is that technology should bring out the best in the technology, and the best in humans. That means robots shouldn't pretend to be humans (like a robot waiter), and that humans shouldn't pretend to be robots (by doing the same difficult task for hours on end, or by performing complicated calculations or search commands).

Case: 'Designers of robots and artificial intelligence should ask themselves what it is they want to optimise in order to improve people's lives. I see a lot of unnecessary technology that was made because the makers think it's cool, but then it just disappears into a drawer after a few years. The question of whether or not it actually helps the consumer is often of secondary importance. Do we really need a speech assistant like Alexa at home? Do we really need a smart refrigerator? Do we really want our children to play with a talking robot doll with cameras for eyes?'

According to Case, robots can do so many more useful tasks than having stilted conversations with humans, and then sending all kinds of data to the manufacturer without the user having any control over it. 'What we need is fewer Silicon Valley utopias, and a lot fewer dystopian visions of robots taking over the world. What we need is a realistic perspective, in which robots do things for humans that humans either can't or don't want to do, or that robots can do better.'

What is the best ratio of robots to humans?

Determining the ideal collaboration between humans and robots is a constant balancing act. That even applies to the most roboticised industry in the world: the automotive industry. You might think that using more robots would automatically lead to more efficient production, but in 2014 the Japanese carmaker Toyota replaced a small number of its robots with humans. Why?

The problem is that robots still can't think about what they are doing, how they are doing it, and how they can improve it. Good craftsmen can. And it was precisely those craftsmen that Toyota noticed were missing when it used too many robots: 96 percent of the entire production process had been roboticised. At Toyota, they put it this way: 'We cannot simply depend on machines that only repeat the same task over and over again. To be the master of the machine, you have to have the knowledge and the skills to teach the machine.'

By reinstating well-trained craftsmen, Toyota was able to reduce waste by 10 percent in the production of crankshafts for its cars. Later, the company managed to cut costs in other parts of the production process as well by replacing a small number of robots with humans. Soon after, the German car manufacturer Mercedes also replaced some of its robots with people.

Toyota's example shows that more robots are not always better, and that it's important to find the right balance, so that what humans and robots do together is better than what either would achieve on their own. Finding that balance is the holy grail of human-robot-cooperation.

In the rubble: the search and rescue robot

Using robots after disasters is literally a matter of life and death. Every year, one million people die from earthquakes, floods, hurricanes, mudslides, mine disasters and industrial catastrophes.

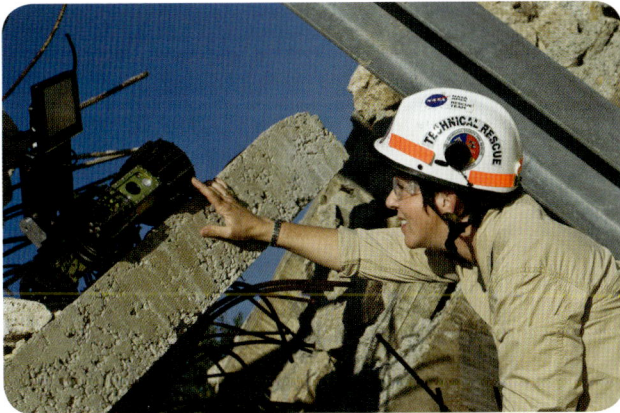

DISASTER: *A robot checking if a building is safe to enter, with Robin Murphy of 'Roboticists Without Borders.'*
Robin Murphy

In 2008, the American Robin Murphy at Texas A&M University founded the programme 'Roboticists Without Borders' in order to train volunteers how to use search and rescue robots like the ones she and her colleagues have developed. Like Leila Takayama, Robin Murphy is also listed as one of the '25 women in robotics you need to know about'.

When a disaster strikes, Roboticists Without Borders can respond with ground robots, flying robots, and underwater robots. Most of them are partly autonomous and partly remote controlled. They are mainly used to help human aid workers to evaluate whether infrastructure like buildings or bridges are safe enough to send humans in to help. Murphy: 'Our robots are the eyes, ears, and hands that humans can use to remotely assess a situation and intervene. And that's important, because these robots help to alleviate the consequences of a disaster so that the local residents can pick up where they left off sooner.'

After the tsunami in Japan in 2011, submersible robots helped

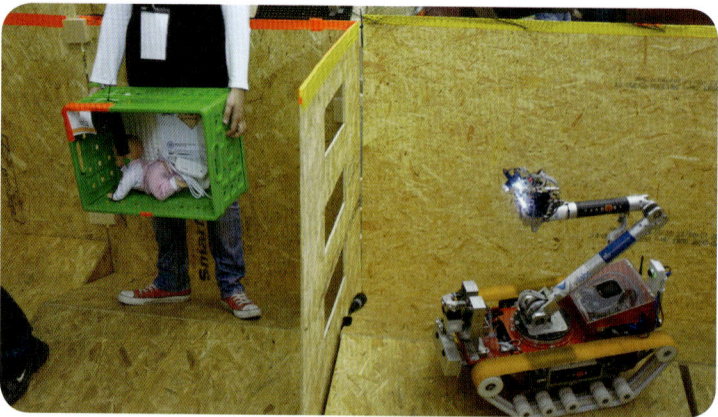

FINDING THE BABY: *A rescue robot takes part in the RoboCup competition.*
Bennie Mols

124

to monitor the damage to the infrastructure along the coast. Thanks to the swimming robots, the work was completed six months earlier than if human divers had to do all of the work themselves.

Murphy explains that the success of her search and rescue robots depends entirely on good collaboration between humans and robots. She has studied the use of search and rescue robots since they were first used in response to the attack on the Twin Towers in New York in 2001, and she has concluded that slightly more than half of all unsuccessful robot missions were due to human error. Murphy: 'After a mine disaster, for example, the robots communicate with the operator by means of a cable. They don't use wireless, because in a collapsed mine wireless communications either don't work or are too unreliable. But sometimes we see that human operators let the robot drive over its own cable, which damages it to the point that the robot can't communicate anymore. Robot designers often expect too much from the robot operators.'

Murphy also concludes that the robot operator should not also be the person who looks for survivors on the screen: 'That's too taxing on the brain. When you have a second person in addition to the operator who focuses solely on looking for survivors, then the chance of finding a survivor is nine times higher.'

The paradox of robotisation: people are needed

One of the aspects that is often overlooked when it comes to robotisation, is that in practice every robot is designed, built, programmed, maintained, and repaired by humans. The robots are also usually under human supervision when they perform their tasks, which means even robotic systems are actually a collaboration between humans and machines. In order for them to function optimally, we therefore need to consider the human factor as well.

In 1983, the English psychologist Lisanne Bainbridge noted in her scientific article 'Ironies of Automation' that increasing automation makes human intervention more crucial when the automatic system does make a mistake. And there is always a chance of that occurring, especially as a result of unexpected circumstances in open systems. Bainbridge's observation is known as 'the paradox of automation'.

Although the aviation industry has become exponentially safer with the application of automatic pilots, we still won't fly in an airplane that doesn't have a human pilot. And that's entirely justified, because the combined system of a human pilot plus an automatic pilot is much safer than either a human pilot or an automatic pilot alone.

When asked whether airplanes will be allowed to fly without a human pilot at some point in the future, Steve Landells, aviation specialist at the British Airline Pilots Association replied: 'We have concerns that in the excitement of this futuristic idea, some may be forgetting the reality of pilotless air travel. However, every single day pilots have to intervene when the automatics don't do what they're supposed to.'

Since the introduction of the automatic pilot in commercial aviation, several accidents and near-accidents have occurred as a result of poor coordination between the human pilot and the automatic pilot. One infamous example is the Turkish Airlines flight to Amsterdam which crashed on 25 February 2009. As the aircraft approached Schiphol Airport, the altimeter suddenly indicated an altitude of minus two metres. The automatic pilot thought that the aircraft had landed, and rapidly decreased the engine power. The human pilots didn't notice the error in time, and the aircraft crashed just short of the runway. Nine people died in the crash, and 120 were injured.

What happens regularly in the aviation industry will also happen on the roads, once our cars and lorries begin to make more decisions for themselves. As in aviation, it's clear that self-driving cars have the potential to cut the number of traffic fatalities, because more than 90 percent of all accidents are caused by human errors that could have been prevented. But self-driving cars will also cause new types of accidents, when the car makes a mistake that the human cannot correct in time.

The cooperation between robots and humans will present new challenges. How will we deal with this shift in responsibilities? How will we deal with the fact that humans tend to trust machines more than their own common sense? How will we deal with the fact that human skills tend to deteriorate when robots take them over? Humans will undoubtedly find solutions to these issues, but the optimal collaboration between humans and robots will not come automatically.

When Isaac Asimov's character Susan Calvin looks back on her work at the end of her life, she says: 'There was a time when humanity faced the universe alone and without a friend. Now he has creatures to help him; stronger creatures than himself, more faithful, more useful, and absolutely devoted to him. Mankind is no longer alone. Have you ever thought of it that way?'

Operating with an intelligent robot

'When I operate using a robot, I take a seat in a kind of cockpit. Through a big pair of goggles, I see the organ that I'll be operating on in three dimensions. It feels like I'm inside the body. My hands hold grippers I use to control the robot, and the instruments at the ends of the robot arms do exactly what my brain wants them to do. Operating with a robot is a magical sensation.'

We spoke with Ivo Broeders, surgeon at the Meander Medical Centre in Amersfoort. In August 2000, he was the first surgeon in the Netherlands to conduct an operation using a robot, and now he wouldn't want to do without. It is not that the robot does anything independently, because the technology is nowhere near that point yet, but as he says: 'The robot is a high-tech instrument that's an extension of the surgeon.'

A surgical robot is one example of a robot that is controlled remotely by a human: a tele-robot. It doesn't make its own decisions, but it's programmable and collects and processes information. The surgical robot refines and processes the information that it receives via the joysticks. That allows the robot to eliminate factors such as the trembling of the human hand. It can also magnify the scale, allowing the surgeon to work down to a square millimetre. Broeders: 'With the robot, I can work more precisely, for example avoiding the tiny nerve endings located around the area of the operation. That causes less damage to the patient.'

Surgery with a robot is the next logical step from minimal invasive operations, or 'keyhole surgery'. Broeders and his team of surgical assistants administer five tiny incisions in the patient's abdomen: four for the three working arms and the robot's camera arm, and one for the surgical assistant responsible for lifting the organ, suctioning blood, or suturing tissue. The coordination between the surgeon, the assistants, and the robot is crucial.

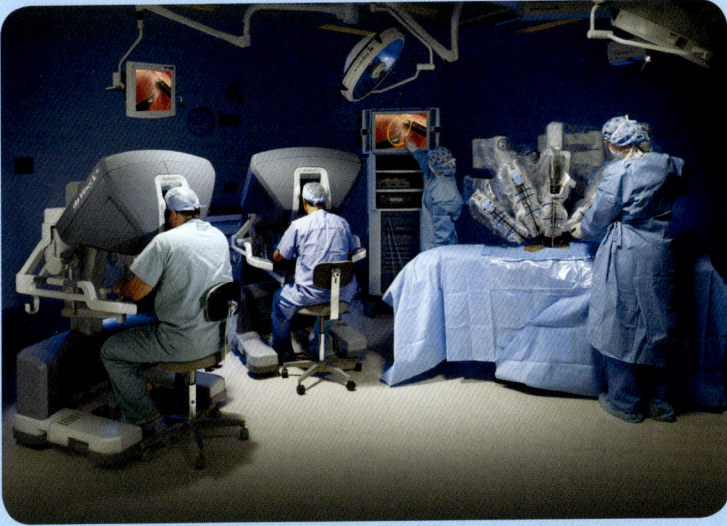

HELPING HAND FOR SURGEONS: *Da Vinci Si robot assists an operation.*

The history of the surgical robot is fascinating. Broeders: 'In the 1980s, the research branch of the US Army dreamt of being able to operate on the wounded on the battlefield by means of a remote-controlled robot. But that dream was completely unrealistic. On the battlefield, you don't have electricity, you don't have time to prepare, and you don't have a flat surface. You need low-tech instruments on the battlefield, not a high-tech surgical robot.'

It was thanks to that dream, however, that the first surgical robot was completed in 1995; not for robotic operations on the battlefield, but for surgery in a hospital. The surgical robot really came to its own between 2005 and 2007, in the field of prostate surgery. Its use spread quickly after that; in 2011, robots performed 90 percent of all prostate operations. In the near future, Broeders thinks that the field of general surgery will become the biggest user of surgical robots: 'Operating on the thyroid, the lungs, the stomach, the large intestine, the stomach lining, the diaphragm, the pancreas, the blood vessels. In fact, anything that's difficult to operate on.'

Surgeons who are already trained to perform keyhole surgeries can learn how to operate a robot in just a few months. Broeders: 'The thing that takes the most getting used to, is that you can't feel for yourself. When you operate, sometimes you have to lift or pull on tissue. With the robot, you can't feel the force anymore, and you have to learn how to operate entirely by eye.'

Passing on that feeling from the surgical robot to the surgeon has proved to be much more difficult than many robot developers had expected. 'That's because of the unimaginable complexity of the sensors in the human body, and the smart connection between the brain and the body. It's also the reason that the dream of a robot that can perform surgery entirely independently will remain a utopia for the next few decades.'

According to Broeders, the realistic future for surgical robots will be a robot that can think along with the surgeon. 'For example, the surgical robot will help me to find structures in the tissue that I can't see on my own. It will also be able to lead me to a specific point much faster. Another development is that the surgical robot will suggest different scenarios for which side is best to approach a specific organ. And finally, the robot will be able to analyse the surgeon's performance. That will allow the surgeon to improve his or her skills. The surgeon of the future will have to control a robot brain with a doctor's heart.'

8 Become a cyborg and walk again

Cyborgs: from science fiction to reality

At the word cyborg, you may think of *RoboCop*, the murdered police officer who is remade as a crime-fighting superhuman with robot parts. Or the Terminator, a robot with a layer of Arnold Schwarzenegger around its metal frame. You may not immediately think of Darth Vader, the villain from *Star Wars*, but he also belongs in this list of cyborgs: after losing his arms and legs in a lightsabre battle, he was fitted with artificial limbs. But not all cyborgs are as threatening as these. Have you ever realised that the animated hero Inspector Gadget, with his go-go-gadget-arms, is also a cyborg?

Cyborgs are humans with mechanical or electronic parts, or robots that are part human, and we generally know them from dystopian fiction made in the 1980s. But when you think about it, cyborgs are all around you. Anyone with a pacemaker, for example, or a cochlear implant: a hearing aid that directly stimulates the auditory nerve. Even an artificial hip or contact lenses are pieces of technology carried inside the body.

Repairing or improving our body using technology isn't such a strange concept. Prostheses are perhaps the most obvious examples. The first prostheses were simply wooden arms or legs. Later on, mechanical joints were developed. Today's prostheses aren't just flexible; some are incredibly intelligent. And they have to be, because it's not easy to make a prosthesis into a fully functioning robotic body part. Take a robotic hand: in order to be as good as a human hand, it has to be strong enough to lift a shopping bag and delicate enough to pick up an egg without breaking it — and of course it needs to decide which of the two to use at the right moment.

Herman van der Kooij, Professor at the universities of Twente and Delft, develops such smart prostheses and robotic aids. His field is 'biomechatronics': a combination of biology, mechanics, and electronics. Van der Kooij: 'In order to make a good prosthesis, you need to build human behaviour, such as walking and keeping your balance, into a robot. So half of my research actually deals with people.'

Van der Kooij started his scientific career with research into how people walk and maintain their balance. 'After a while, I thought: all these models are fun, but I need an application. I wanted to use my knowledge about human movement to be able to help people. That's why I started working on wearable robotics. Prostheses today are actually a type of robot, because they have motors, sensors, and intelligence.' Van der Kooij develops many of the motors and sensors himself, to stretch his skills.

Exoskeletons: high-tech walking frames

The American Hugh Herr was only 18 years old when he lost both of his legs in a climbing accident in 1982. 'My line of reasoning is that a human can never be "broken",' he explains during a TED presentation. 'Technology is broken. Technology is inadequate.' With that attitude, he went to work building a piece of technology that would enable him to walk and climb again. Within six months, he was able to return to rock and ice climbing. He quickly realised that he could cleverly adapt to different climbing circumstances by developing prostheses with unique characteristics, such as tiny 'feet' that allowed him to stand on narrow ledges. With his special-ised prostheses, he was eventually able to climb better than he could before his accident.

Herr: 'I imagined a future where technology was so advanced that it could rid the world of disability. A world in which neural implants would allow the visually-impaired to see. A world in which the paralysed could walk, via body exoskeletons.' Today, 35 years after his accident, he is head of the Biomechatronics group at the Massachusetts Institute of Technology, where he helps to build that future himself. His developments include a robotic prosthesis for a dancer who lost her lower leg in the bomb attack at the Boston marathon in 2013. The prosthesis helped her to learn how to walk again, and even dance. At a TED conference a year after the attack, she performed with her prosthetic leg for the first time in front of an audience.

Walking and moving aren't just thinks that make us human; they are also healthy. Or rather: not walking is unhealthy. That doesn't only apply to people who cannot walk due to paralysis or an amputation, but also to people who have trouble walking as they get older. Van der Kooij: 'It's important for people to keep walking. People who sit in a wheelchair often develop other

STEPPING INTO THE FUTURE: *Dancer Adrienne Haslet-Davis (left) performs for the first time with her prosthetic leg, made by Hugh Herr (right).*
Steve Jurvetson

complaints: pain, deteriorating bladder or intestinal function, osteoporosis... All of those complaints can be reduced if people have the opportunity to walk every once in a while, for example with the help of an exoskeleton.'

An exoskeleton is a kind of robotic suit that can help paralysed people walk again. For people who need a wheelchair to get around, an exoskeleton can make a big difference; not only because it allows them to walk again, but also because they can stand and talk to others at eye level. Van der Kooij: 'There are experiments you can use to test how long people intuitively think that their arms and legs are. People in a wheelchair draw their legs as shorter than they really are. After walking in an exoskeleton for a while, they develop a more complete image of their own bodies.'

It is already possible to walk in an exoskeleton, but there is still plenty of room for improvement. 'For most exoskeletons on the market, every step is the same. You make a movement with your torso, or you shift your weight in a certain way, and the exoskeleton takes a step. The step is the same every time; you can't take longer or shorter strides. You often can't climb the stairs either, or walk over rough terrain with height differences, for example. And it's extremely tiring, because you have to think about every step, and you need crutches to maintain your balance.'

How would he like to improve them? 'Actually, I'd like to build biological characteristics into the devices, some kind of muscle characteristics and reflexes. That would make walking in an exoskeleton easier and more natural.'

Exoskeletons and other robotic aids can also play an important role in preventing injuries or conditions caused by hard physical labour. Van der Kooij therefore also studies exoskeletons that support the body by taking weight off of the back or shoulders. The bar for that work is set very high: 'Experience has shown that workers won't use something like that if it's even the slightest bit cumbersome. That applies to conventional aids as well: occupational health and safety norms require all sorts of equipment, but

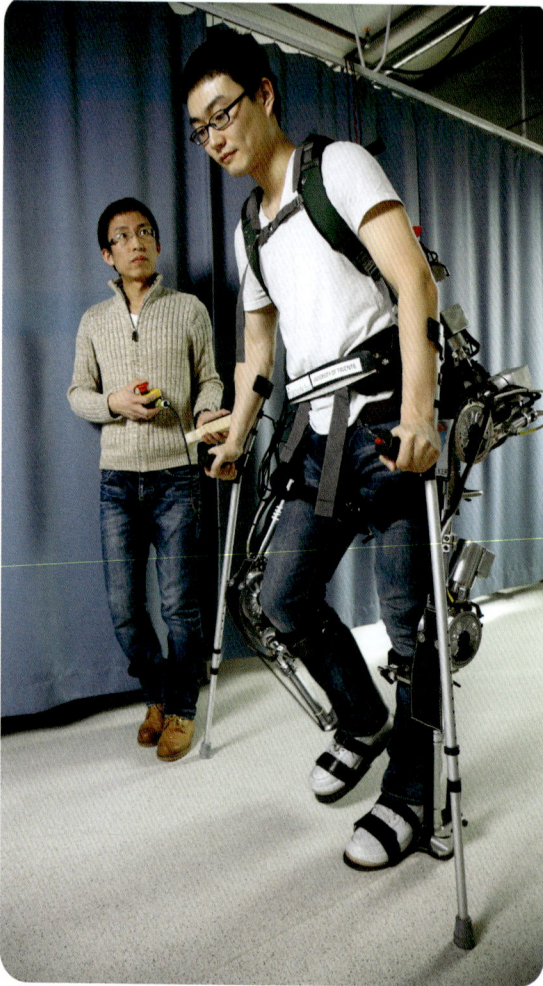

PROTECTING HUMAN JOINTS: *Exoskeletons can help people walk again after accidents.*
Herman van der Kooij

workers often don't use it, because it's such a bother.' Think of back harnesses to reduce the load on the lower back.

Van der Kooij is working on a kind of 'exosuit' that's comfortable and easy to use for people who do physical work. The exosuit is a kind of robotic trousers that is not rigid and heavy like an exoskeleton, but still provides enough support. 'The main application of these kinds of aids will be for prevention. That applies for builders, but also for people working in health care. And surgeons, because they are constantly bent over the patient during an operation. We call that 'human-centred robotics'. So not robots that replace humans, but robots that help and support them.'

Robotic parts must become a part of you

In order to get the feeling that a prosthetic limb is part of your own body, it is important that the prosthesis can pass on the right sensory information. You can observe that when you see the 'rubber hand illusion'. In this illusion, you sit at a table holding a screen. You put your left hand out of sight behind the screen, and a rubber left hand is placed in clear view. If someone then brushes your left hand and the fake left hand at the same time, your brain thinks that the rubber hand feels like your own hand. The feeling is so strong that you will reflexively pull back your own hand if someone threatens to prick the rubber hand with a needle.

Building senses into the prosthesis will therefore make it feel more like it's part of your own body. Unfortunately, it's still very difficult to do that well. Van der Kooij: 'Sensors often use a lot of energy. Each sensor consumes a few Watts of electricity. Humans are different: we have a lot of senses, but they all use very little energy. Many human senses only send a signal at the moment it is absolutely necessary, such as when something changes. Sensors are always on, so they are always consuming energy.'

Good sensory feedback is difficult to achieve, but it's necessary in order to be able to control a prosthesis without thinking about it. 'With most of today's prostheses, you have to look at them while you control them. That's because you can't feel if you're touching or holding something. Right now, researchers are working on a prosthetic hand that is connected to electrodes in the brain, which allows you to not only control the hand, but also feel temperature or pressure. If that works, then you'll have situation that is almost normal; a mirror image of a natural limb. At the moment, it's still difficult and expensive, but in 20 years I think that those kinds of products will be available at an affordable price.'

In 2016, the American defence research institute DARPA built the very first prosthetic hand that can be controlled by the brain, as well as send back sensory information. The brain interprets the electric currents that the robot hand sends in the same way that it interprets the information from natural senses. Nathan Copeland, who has been fully paralysed from his shoulders down since a car accident in 2004, tested the prosthesis. A brain implant allowed Copeland not only to control the robot hand, but also to feel with it. Research Leader Justin Sanchez: 'At a certain moment, the team decided to touch two fingers instead of just one, without telling him about it first.' Copeland noticed it immediately. Sanchez: 'That's when we knew that the feeling he had in the robot hand was virtually natural.'

Van der Kooij predicts that in the more distant future, over 50 years or so, mechanical exoskeletons and prostheses may be unnecessary altogether for people with paraplegia. Paraplegia is a spinal injury in which the signals from the brain cannot reach the legs. The leg muscles still work, but the brain can't control them. 'If you insert stimulators in specific places in the nervous system, then send electric currents through them, you can make the muscles move.' That makes it possible to bypass the paraplegia. 'In fact, you can control the body just like it was a robot. One nice side effect is that moving your own muscles can also initiate all sorts of recovery processes in the muscles.'

That may sound like science fiction, but researchers have already begun work on the first experiments. Van der Kooij: 'It's already possible in insects. Their nervous systems are simple enough for that. You can stick a few electrodes in a bee, and then you can control it as if it were a kind of biological drone. So the technology is already under development. You'll be able to let paralysed people walk again, without needing cumbersome mechanical components. Hardware is increasingly merging with 'wetware' — the human body.'

Soft robots filled with liquid or air

Another way to have robotic parts feel more lifelike is through soft robotics: the use of soft materials such as flexible polymers or bags filled with fluid or air. Baymax, the big robot from the film *Big Hero 6* (2014), is inflatable. With its soft body, Baymax is not only extremely cuddly, but also as flexible as a balloon animal.

Soft robots have all kinds of advantages over traditional rigid robots. They can change shape to move through tiny openings, and they can move more easily over soft surfaces, such as sand or mud, without sinking. Soft, flexible robots and prostheses are also safer for humans. Miniature soft robots may also be safer for medical applications in which they move inside the human body. Materials that can turn hard or soft on command are ideal for use in the robot suits that Herman van der Kooij envisions, because they behave more or less like muscles. People with a foot drop, for example, can be helped with a kind of sock that stiffens as they lift their foot, then relaxes at the moment the foot is set down again. The first pneumatic muscle was developed as early as the 1950s: it was basically a balloon in a sleeve woven from non-elastic fibres. When the balloon was inflated, the fibres expanded, making the sleeve shorter. Unfortunately, the pneumatic muscle is nowhere near as efficient as a human muscle.

Soft robots are also useful for applications that are unrelated to dealing with humans. Researchers at Standford University have created a kind of inflatable snake, which can change direction as it is inflated, allowing it to navigate around corners. This 'robot snake' can be used to extinguish fires in difficult-to-reach places, or to move through rubble to explore the interior of a collapsed building. When combined with sensors, the snake can also move autonomously towards light or sound.

Soft robots are already being used for industrial applications. The company Soft Robotics Inc. develops soft robot grippers that can grasp tomatoes, for example, and sort them according to ripeness without damaging them. Since they are made of flexible materials and controlled pneumatically, the robot grippers can safely grasp delicate objects in all shapes and sizes; from apples to raw eggs, and from stuffed animals to cream puffs.

The robots built by Soft Robotics have soft hands, but the rest of the robot is made of rigid materials, such as metal and plastic. It's difficult to make a robot that is entirely soft, because components such as batteries and motors are still all made of hard materials. That means 'soft' robots are generally only partly soft.

However, the first fully soft robot was created in 2016: Octobot, a robot in the shape of an octopus. The flexibility of the octopus was a major source of inspiration for the researchers at Harvard University: an adult octopus can press itself through a hole or tube measuring just a few centimetres in diameter. Instead of a traditional motor, Octobot has a pneumatic system without hard components: combining two substances produces a gas that inflates Octobot's arms. That makes Octobot a completely 'soft' robot. At the moment, Octobot can only move its arms up and down, but the researchers hope to be able to make it crawl and swim in the near future.

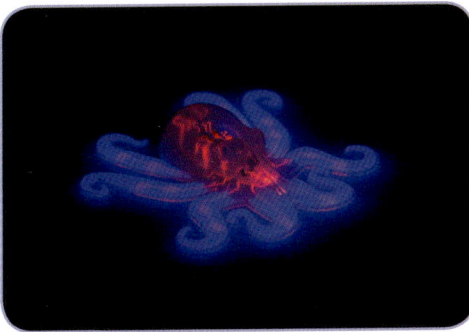

SOFT ROBOTS:
Octobot is the first robot made wholly from soft materials.
Lori Sanders/Harvard University

BEAM robots without a brain

Surprisingly, there are also autonomous robots without a brain. These so-called BEAM robots — for Biology, Electronics, Aesthetics and Mechanics — can navigate based on reflexes alone. The robots have wheels that are connected directly to sensors, allowing them to navigate towards — or away from — a source of light or sound. These robots basically skip the middle step of the sense-plan-act cycle.

That may sound strange, but some animals don't have a brain either: starfish and sea cucumbers live their entire lives with just a nervous system and some sensors. Having a brain consumes a lot of energy — up to a fifth of the energy used by the entire body in humans — so if an organism or robot can deal with a specific ecosystem without one, it's an efficient solution.

By developing new materials that can adapt their characteristics to their surroundings, robots will be able to do even more for us in the future. In the first decades of robotics, robots were machines that were placed behind fences in a factory, outside of the reach of humans. Then, in the 1980s, mobile robots gradually began to appear that could move around safely in the same space as humans. In the early years of this century, robots could deal with humans at a basic social level. And with the invention of robotic prosthesis, exoskeletons, and soft robotics, robots are literally creeping into our skin. Over time, robots have gradually become closer to humans.

Paraplegia — never give up

In 2004, Claudia Bosch-Commijs suffered a spinal injury after falling from her horse. 'I'll never be able to walk again', is what she told her friends at the time. She was right too — for 12 years, until she stood in an exoskeleton for the first time in 2016. She noticed immediately that she had never really become accustomed to her wheelchair. 'I had a very emotional reaction. I hadn't expected that, because I had been in a wheelchair for so long. But standing felt so familiar that I thought: I'll never really get used to sitting down. You try to ignore it a bit when you're in a wheelchair, but when you stand up again and look at people eye-to-eye... As bizarre as it sounds, it turns out to be so important.'

The moment was recorded in the Dutch documentary Sta op en loop ('Get Up and Walk'). At the time, Commijs was participating in a study at the Sint Maartenskliniek in Nijmegen, where she was fitted with a commercially available Rewalk exoskeleton. She was also participating in Project March at Delft University of Technology at the time, where a team of students was designing a new exoskeleton with even more functionalities.

There were several differences between the exoskeletons, Commijs says. 'With the Rewalk, you can't take steps yourself, you have to move along with the exoskeleton. You also can't step backwards or sideways.' That makes the Rewalk less suitable for daily use. 'If you don't reach the kitchen perfectly, you have to shuffle forwards for the last bit.' The exoskeleton made by the students in Delft is still under development, but thanks to an extra pair of hip joints, it will eventually allow the user to take steps diagonally and sideways. 'Plus, you'll be able to decide how big the steps should be. That means you will walk the suit, and with the Rewalk it's the other way around.'

Neither the Rewalk, nor the March can maintain their balance yet, so the user still has to walk with crutches. Commijs: 'We already have robots that can balance well, but it's a bit more difficult for an exoskeleton holding a person. Try to tell a robot if a movement was intentional or unintentional. Our brain registers that immediately. That's when you notice: the human body is truly ingenious, and it's not easy to imitate it. But that would be a major area of improvement. If you can walk without crutches, then you have your hands free for other things.'

Anything else? 'It could be a bit slimmer', Commijs adds. 'The current exoskeletons don't let you sit in your wheelchair or in the

REBUILDING A BODY:
After four weeks in an exoskeleton, Claudia Bosch Commijs could halve her medication.

car. If that were possible, it would be easier to just get your crutches and stand up for a bit.'

Commijs still noticed many benefits to walking with an exoskeleton. 'It's already great to be able to walk with paraplegia, and to be able to talk to people at eye level. And it's also very good for your health. With my nerve pain, I can't function without medication. But after four weeks of walking in an exoskeleton, I could cut the medication in half, and after six weeks I was able to stop completely. The circulation in my legs is better, I have less back pain, and the longer you use it, the denser your bones become.' The health benefits were noticeable after walking for only an hour in the Rewalk three times per week. 'It's more like fitness than walking, especially in the beginning, when you still have to learn how to use it.'

After the study, Commijs had to return her exoskeleton, so she decided to launch a crowdfunding campaign to buy her own Rewalk. In the summer of 2017, she finally hit the target of 87,000 euros, and when we spoke to her, she had just taken delivery of her robotic suit. Hopefully, she will be able to start walking in it within a week. Commijs looks forward to being able to use the exoskeleton again. 'I want to take the dog for a good long walk. And I'll absolutely bring it with me to parties. Then I can stand at table and drink a beer, just like everyone else.'

9 Evolution designs the best robots

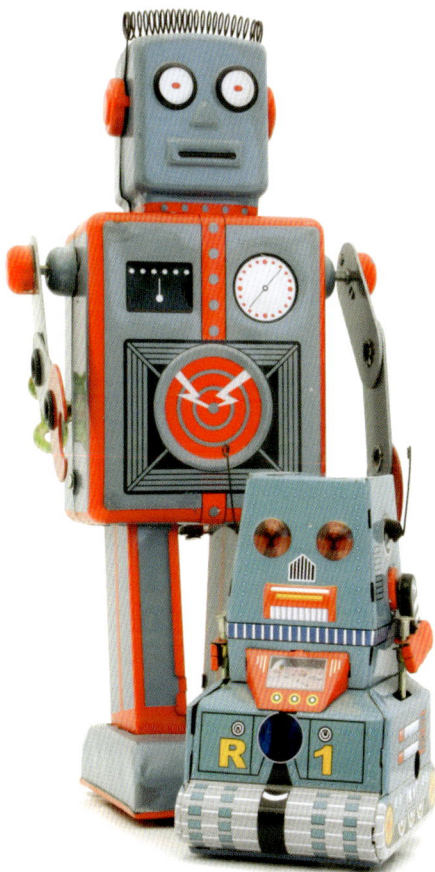

How robots travel in a bumpy world

Most mobile robots move around on wheels. Vacuum cleaner robots do it, self-driving cars do it, the delivery robot in a hospital or a hotel does it, and even the social robot Pepper uses wheels to get around. Wheels are perfect for moving around quickly and efficiently on a flat surface: on motorways and factory floors, in the office, and at home.

But take a look around in the animal kingdom. Can you think of an animal that uses wheels? No, not a single one. Humans invented the wheel, not nature. Before humans arrived on the scene and started building roads, the surface of the earth was generally uneven. Wheels aren't very handy for travelling over rough terrain, so organisms evolved other ways to move around: snakes can slither through narrow openings, but the vast majority of animals and insects evolved legs: from creeping centipedes and jumping kangaroos, to four-footed dogs and the erect, bipedal humans.

Of all of these species, one is the undisputed world champion of climbing walls: the gecko. The gecko is a reptile with two large eyes and four legs, each of which has five fingers. It can walk upside-down along a ceiling, in the absolute conviction that it will never fall down. In 2000, biologist Robert Full at the University of California Berkeley was the first to find the answer to the question of how the gecko acquired its brilliant climbing skills.

For decades, Full has studied how crabs, ants, beetles, cockroaches, and geckos perform seemingly impossible antics in order to reach the most challenging of places. He discovered that the cockroach can extend its exoskeleton in order to fit into a crack that is four times narrower than the creature itself. He also discovered how the gecko's fingers can quickly stick to a surface, and then let go again just as quickly.

Each gecko foot has half a million tiny hairs (30 to 130 thousandths of a millimetre long, and 5 thousandths of a millimetre wide). Each hair has even thinner hairs protruding from the end (only 0.2 thousandths of a millimetre wide), each of which has a spade-like tip. These tiny hairs can make contact with a wall, ceiling, or any other surface. Full discovered that the spade tips are so small that the molecules that make up the hair and the wall are actually attracted to one another. The combined attractive force of all of those tiny hairs is so powerful that the gecko can 'stick' to any surface, even when it's hanging upside down. And to release itself from the surface, all the gecko has to do is curl its fingers.

Although Full is a biologist, he has had a major influence on robotics. He works together with roboticists, giving them new ideas for the designs for their robots, because there is no better source of inspiration for robots that can walk across rough surfaces, climb walls, or crawl through tight spaces than nature. 'Use nature for inspiration, and then build things that don't exist yet', he says in the book *Robo Sapiens*.

The idea isn't to copy nature in every detail — nature is too messy and imperfect for that — but rather to understand the fundamental principles and apply them to the field of engineering. He suggests that humans use nature's examples for the things that have human applications, and leave out the things we don't need.

Full's discovery of the gecko's secret has inspired roboticists to build gecko robots, with names like Mecho-Gecko and Sticky-Bot. These gecko robots use a kind of gecko-tape developed by Nobel Prize winner Andre Geim in 2003. Although the gecko tape is made from a different material than the hairs of the gecko's foot, it shares the same structure and sticks to surfaces using the same principle. Gecko robots also mirror the peeling motion that geckos use to release their feet from a wall.

Gekko robots are still in the process of development, but in the future they may be able to clamber up the smooth surfaces of solar panels in order to inspect or repair them. NASA is also developing its own gecko robots. The engineers at the American space agency think that they will be able to climb along the hull of a space ship, sticking perfectly to the surface to conduct inspections or repairs without a care for the near-vacuum of space, the lack of oxygen, or weightlessness. Gecko robots may also be able to look for signs of life on the rough surface of Mars. On Earth, they could serve as search and rescue robots to look for survivors in the rubble after a disaster.

Robothand has nature's grip

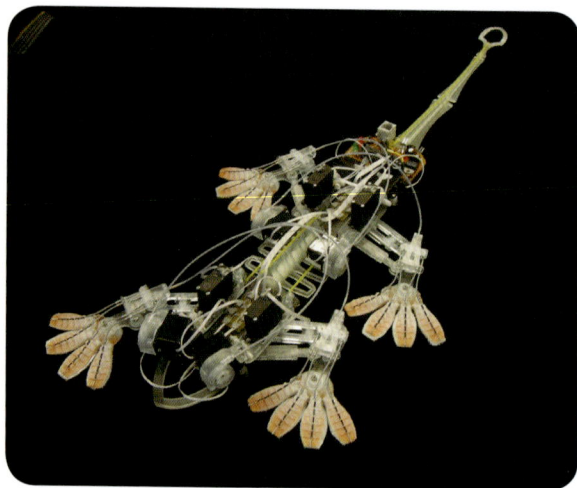

JUST LIKE A GECKO: *The inspiration for this Stickybot robot lies in nature.*
National Science Foundation

Professor of Biorobotics Martijn Wisse at Delft University of Technology also finds inspiration in nature. In 2005, he and his American colleagues built a two-legged robot that walks almost as efficiently as a human. In the five years that followed, he developed a robot hand inspired by the human hand. This robot hand can pick up pieces of fruit and vegetables without damaging them, something that traditional robot hands cannot do.

Both robot applications are the result of close observations of how people walk and grasp, without trying to make an exact copy. 'It's never good to blindly copy nature', Wisse says. An airplane doesn't fly the same way as a bird, which is why a Boeing 747 can carry hundreds of people, and do it faster than even the fastest bird. 'Still, we robot builders are always a bit jealous when we look at nature. Compared to robots, animals can do a phenomenal number of things: they move so smoothly, react so quickly and powerfully, and can jump so accurately. There is so much variation in nature, and it all seems to be so easy.'

Like the father of robotics, Joseph Engelberger, Wisse became fascinated by robotics after reading *I, Robot* by Isaac Asimov. 'I read it as a teenager, and was immediately gripped by the fact that the robots in the stories could do so much useful work for humans. I thought it would be fantastic for robots to prepare your food, clean your house, or even build a house for you. In principle, you could also have robots build other robots. When I look around the world today, it's a bit disappointing what robots can do. So much manual labour is still done by humans. That's a waste of time, and of manpower. Robots still have a world to win, and I want to contribute to that.'

With the development of robot grippers for the fruit and vegetable industry, Wisse has already achieved that goal. Until recently, humans had to sort the green, red, or yellow bell peppers or pack chicory in a crate by hand, one piece per second, eight hours per day. One third of the cost of production in the fruit and vegetable sector is human labour. Now, Wisse's robot hands can take over that work.

What's the secret to these robot hands? Wisse: 'In short, it comes down to throwing overboard the approach that roboticists followed for decades. Robots have traditionally been machines with motors that move arms or legs. The robot measures the position of each joint, and then the computer determines how to fine-tune the settings in order to bring the arm or leg into the right position. That works perfectly for a welding robot in the automotive industry, but it's completely wrong for the fine finger work that humans can do. It's especially noticeable for objects that you don't know the exact size of in advance, like pieces of fruit or vegetables.'

Initially, roboticists had hoped extra sensors would accurately measure the force on the robot fingers. But that required adding more motors to regulate the position of the fingers, which was cumbersome and still didn't work well. Wisse: 'That's when we took a good look at how the human hand works when it grasps an object. My colleagues were developing prosthetic hands, and they came across something unusual. Our fingers have fewer muscles than joints. Try to move the three knuckles of your fingers separately from one another. It's impossible. When you try to move the first knuckle, the second one moves along with it. A robot builder would never think of that, because it means you can't control each knuckle independently.'

Is that a fault in the design? Wisse says no. 'Since each finger on our hands uses two tendons for three joints, you can move the finger joints enough in relation to one another, while at the same time the force is always properly distributed. When you close your fingers around a round object, your fingers take on a round form. But when you close your fingers around a book, your fingers take on a flatter form. It's not because you're so intelligent, and the force sensors in your skin feel everything, while your brain calculates that your fingers should take on a specific shape. That is far too complicated. Nature has developed a design that solves the gripping problem in a purely mechanical way, without using observation and intelligence.'

Once roboticists understood how the mechanism works in a human hand, they were able to copy it in a robot hand fairly easily. In 2010, Wisse was awarded a patent for a specific design for a robot hand that's ideal for the horticultural industry. His robot gripper can deal with a wide variety of shapes, sizes, and textures in pieces of fruit and vegetables. That same year, the robot grippers were brought to market by the firm Lacquey, which he founded together with his colleague, Richard van der Linde.

SORTING PEPPERS: *Robots can be trained to do mundane tasks, freeing up humans for more interesting jobs.*
FTNON Lacquey

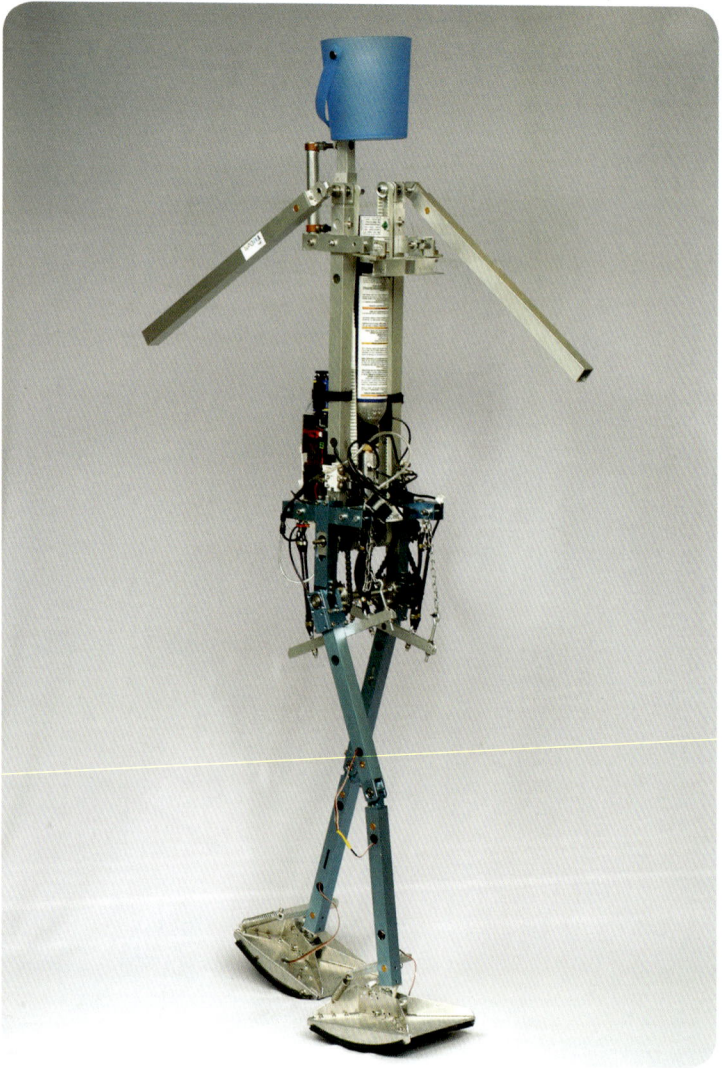

WALKING IS TRICKY: *But Denise has mastered putting one metal foot in front of another.*
Martijn Wisse/TU Delft

A two-legged walking robot

For decades, roboticists tried to use precise measurements and adjustments of the angle of hips, knees, and ankles to construct two-legged walking robots. Over time, robot motors became more powerful and computing capacity rocketed, but still walking robots were so clumsy in their movements that it put Mother Nature to shame. Wisse: 'Here, too, there was too much focus on imitating how the walking movement looked, instead of trying to discover the most important underlying mechanical principle.'

The breakthrough came when the American researcher Tad McGeer discovered that you can describe the way humans walk as two pendulums connected at the hips. Wisse: 'You can see the standing leg as a single pendulum, because the knee is more or less rigid. The swinging leg, on the other hand, is like two pendulums that move in relation to one another at the knee joint. Based on that insight, McGeer built walking robots that can walk down a slight incline in a very natural way on their own, without a source of power.'

Together with researchers at Cornell University in America, Wisse adapted McGeer's design and added a small motor to the robot. The result was the creation of the walking robot Denise in 2005. 'Denise's motor gave a little push for every step. That way, she didn't need an inclined slope anymore, and she was able to walk over a flat surface in a natural way with very little energy. Denise's walking movements are almost as efficient as those of a human. That's because she doesn't have to do much: she doesn't measure, she doesn't calculate, and she doesn't adjust. She just lets gravity do its job.'

What is true for humans is also true for simple insects. Biologist Robert Full discovered that insects like the cockroach are able to stabilise themselves while walking without using their tiny brain. Instead, their mechanical structure does all of the work. 'It's as if their springy legs do the calculations themselves; as if the control algorithm is built into the creature's body.'

Wisse's walking robot is much less useful than his robot gripper, however. 'The walking robot is only stable on a smooth floor surface, where there's no wind or other disturbances. If the robot is disturbed in any way, she falls over.'

That brings him to the biggest challenge for mobile robots: the development of a robotic equivalent for muscles. 'Our robot is very efficient, but she has very little strength. Walking robots that have a lot of power and that are stable when they walk, like the four-footed Big Dog by Boston Dynamics, consume a lot of energy. When you kick Big Dog while it's walking, it stays on its feet, but that costs a lot of energy. The field of robotics still hasn't managed to combine efficiency and power. Muscles are perfect for that. With a human muscle, when you don't need it, it doesn't do anything. But when you need it, it's extremely fast and powerful. It's flexible, it's elastic, and it can provide enormous amounts of force at low speeds. We still don't have technology with that combination of characteristics.'

That is precisely why two-legged robots make us laugh when they have to walk around in the real world, instead of the lab. The internet has countless videos of humanoid robots that can't break their fall when they tumble over for the silliest of reasons.

One major difference between robotics and biology involves the characteristics of the materials. Robotics builds machines out of strong, durable, hard, and lifeless materials. They can go for years with only the most basic care and maintenance. Robotics also uses relatively large components, often with right angles, and with relatively few motors and sensors.

In the field of biology, on the other hand, you see tiny parts, bent, flexible structures, and a lot of muscles and sensors. Biology uses soft materials that are constantly growing and being replaced; living tissue. 'Every individual atom in your body has probably been replaced by a new one at some point', Wisse says. 'That process of constant replacement, growth, and repair is necessary to keep biological materials operational. The field of robotics hasn't even started researching materials that can do the same in a robot.'

BREEDING ROBOTS:
The world's first robot baby (foreground) and its parents: father (blue) and mother (green).
Guszti Eiben

The first robot baby

The most important difference between robotics and biology is, of course, the fact that robots are lifeless machines, and organisms are living machines (with apologies to those who have difficulty seeing living beings as machines). Plants and animals reproduce themselves generation after generation according to the Darwinian principles of variation, selection, and heredity.

Could we build robots that reproduce in a similar way, and that adapt to their environment from generation to generation? The Hungarian Professor of Artificial Intelligence, Guszti Eiben, is looking for an answer to that question in his work at the Free University Amsterdam.

In its robot laboratory, he shows us the world's first robot baby, which was born there in May 2016. The robot baby looks more like an insect than an infant. It is made up of five green blocks, two blue ones, and a white one, which are connected by joints. The baby uses tiny motors to move around and has a Raspberry Pi computer for a brain. The robot's father is made of only blue blocks, and looks like a spider, while the mother robot is made of only green blocks, and looks like a gecko.

Eiben explains how the first robot baby was conceived: 'We start with a father and mother robot, that are each programmed with their own genetic code. Next, the parent robots both learn their own methods for walking, since they each have different forms. Once they can walk well enough, we consider them to be 'sexually mature'. They are equipped with light sensors that they use to detect the mating corner we built for them, and which we have marked with a red light. Since we're in Amsterdam, it didn't take long for people to joke that we'd built a kind of Red Light District for robots. Both of the robots then crawl towards the corner in their own way.'

Once the father and mother robots come close enough (but without making physical contact), they each send a WiFi signal with their genetic code to a computer located elsewhere in the lab. The computer then crosses the two genetic codes in a manner similar to the way sexual reproduction works in living organisms. The baby robot's new genetic code determines how many green and blue blocks it is made of, for example. But like in the natural world, random mutations can also occur. Eiben: 'In the middle of the baby robot, we can see a white block. But neither the father nor the mother used a single white block in their construction. We also didn't know what the baby robot would look like in advance.'

Next, the computer sends print commands to a 3D printer based on the robot baby's genetic code, and the printer produces blue, green, and white blocks one by one. These are the basic building blocks for the baby robot. Eiben: 'At the moment, we can't print out all of the components for the baby robot, such as the wires, LED lights, sensors, and computer chip. And the components can't assemble themselves, either. We use a PhD student for that, because it's by far the cheapest solution.'

The parent robots don't exchange physical genetic material, the mother doesn't provide the material that the baby robot is made of, and none of the three robots are self-sufficient when it comes to energy. If the battery is empty, then the PhD student has to recharge it. There are still big differences between robotics and biology, but as a proof-of-concept, the first robot baby was a successful experiment.

Eiben admits that he is not a roboticist, and that roboticists don't consider him to be one of their own. 'I'm a Professor of Artificial Intelligence, and what I do is apply artificial intelligence in a body, what we call 'embodied artificial intelligence'. My position is: matter matters! We know that in biology, mind and body evolved in interaction with one another. And that is precisely what I want to study with my robots: how do the robot's hardware and its software evolve in interaction with each other?'

Working out the best path through evolution

With enough research funding, Guszti Eiben would like to subject thousands of robots to the principles of Darwinian evolution at the same time, in order to study which bodies develop in certain ecological niches, and what influence the differences between the bodies have on the evolution of each robot's brain. Eiben calls such a laboratory environment an 'EvoSphere'.

'An EvoSphere is a hardware model for evolution. I don't necessarily want to make living robots. It's rather that I want to make robotic systems that have a number of interesting characteristics of real life. Robots are easier to observe and program than animals. We can also constantly read the robot's sensors and brain, without having to ask permission from an ethical committee. At least, not yet...'

Does Eiben think that there is a fundamental difference between robotics and biology? 'That depends on what day you ask me', he says. 'Today I think that there is a fundamental difference, but there are also days when I don't. There are good arguments for answering either yes or no. People who don't see a difference claim that it's about the principles of evolution, and not about the substrata that those principles act upon. On the other hand, we add a lot of things to our simple robots as we go, while nature builds everything from the ground up. And nature uses molecules as the building blocks, where we use actual blocks.'

In addition to being fundamental research, Eiben considers his work to be a contribution to a future technology for making new types of robot bodies. 'When we want to send robots to places that we aren't familiar with, it's difficult to come up with the optimal design in advance. Say that we want to send evolutionary robots to Mars, Antarctica, or the tropical rain forests; could robot evolution find the type of robot body that works best in that environment on its own? Does it need wheels or legs? And how many legs? What kind of legs? What kind of cognition should the robot brain have? Evolution is the best designer we know of.'

Robotic falcons can chase away the birds

Edmonton International Airport in Canada is the first international airport in the world to use a robot bird of prey to scare off birds. The Robird is made by the Dutch firm Clear Flight Solutions in Enschede, which has been developing it for commercial use since late 2012. The company is a spin-off of the robotics research conducted by Professor Stefano Stramigioli and the aerodynamics research by Professor Harry Hoeijmakers, both from the University of Twente.

Wessel Straatman is an R&D engineer at Clear Flight Solutions, and has been involved in the technical development of the Robird since the company was founded. 'The Robird looks like a falcon and flies like a falcon. It works by using other birds' primal instinct to try to escape from their natural predators.'

PREDATOR: *Robird mimics the flight of a real falcon to scare away birds from Edmonton International Airport.*
Clear Flight Solutions

According to Straatman, bird strikes at airports are estimated to cost the commercial aviation industry $3m - $8m each year. 'That's mainly due to repairing damage, such as when birds get sucked into jet engines. But there is also the secondary damage, when aircraft have to be inspected after a bird strike. That often causes flight delays.'

The Robird is a remote-controlled robot falcon with a wing span of one metre and a weight of 750 grams. It looks like a real falcon in every detail: size, shape, colour, flight speed, even how it beats its wings. It doesn't fly autonomously, however; it's controlled by an experienced RC airplane pilot. Straatman: 'Almost anyone can fly an ordinary drone, like a quadcopter, but only the most experienced model airplane pilots can fly a Robird, which flies by beating its wings. I've tried to fly one myself, but I couldn't do it.'

The model aircraft pilots that the company hires as drone pilots first receive intensive training from a falconer on how a real falcon hunts its prey. 'Because it's crucial that the Robird displays the realistic behaviour of a real falcon. In less than five minutes, it has to act like it's flying aggressively and attack from above. Birds often form groups at the airport, so our drone pilots can use the Robird to drive them away from the runway like a herd of sheep.'

For years, airports have used all sorts of technologies to chase away pesky birds: from speaker systems to light guns, and from windmills to off-road vehicles. Straatman: 'Unfortunately, birds soon become accustomed to technological solutions. Birds are much smarter than you might think. But they'll never get used to their natural predators. And the birds can't tell the difference between our Robird and a real falcon.'

In addition to scaring birds away from airports, the Robird can also help reduce bird-related problems in agriculture and waste processing. Tests with the robot falcon have already been conducted in both sectors. Straatman: 'At a waste processing facility, the Robird could reduce the number of birds by 70 to 90 percent, depending on the species of bird. At a blueberry farm, we were able to limit the lost harvest from 25 percent per year down to one percent.'

The current version of the Robird even has an automatic pilot that allows it to fly in a straight line when the drone pilot lets go of the controls. It also uses a GPS receiver to make sure it doesn't fly over the runway. 'In the future, the Robird will become even more autonomous. It'll be equipped with cameras, so it will be able to see where the birds are located and where it is flying. It will be able to think about how it should fly to chase away birds, and it will even be able to adjust its plans mid-flight. Eventually, it should be able to land on its own in order to recharge its battery. All of that is technically possible today, but at the moment the legislation and regulations for flying drones are holding back those developments.'

Interestingly enough, in the world of drones the reverse is happening: real birds of prey are being trained to pluck unwelcome drones out of the sky at major events.

As well as chasing birds from airports, Robird can also reduce bird nuisance in agriculture and horticulture and even rubbish dumps. Straatman: 'At a waste plant, the Robird could reduce the number of birds by 70-90 per cent, depending on the bird species. For a farmer who grows blueberries, we could reduce the annual harvest loss from 25 per cent to 1 per cent.'

The current Robird even has an autopilot that keeps it moving forward when the drone pilot relinquishes control. It uses a GPS to prevent it from flying into unauthorised areas, such as above a runway. 'In the future the Robird will get more and more independent. It will get several cameras so it can see where the birds are and where it flies. It decides how to fly to chase away birds and can even change its tactics in mid-flight. Finally, it should also be able to land independently to recharge its battery. From a technical point of view, all that is possible, but at the moment it's forbidden by the laws and regulations on drones.'

10 Swarming robots show the wisdom of crowds

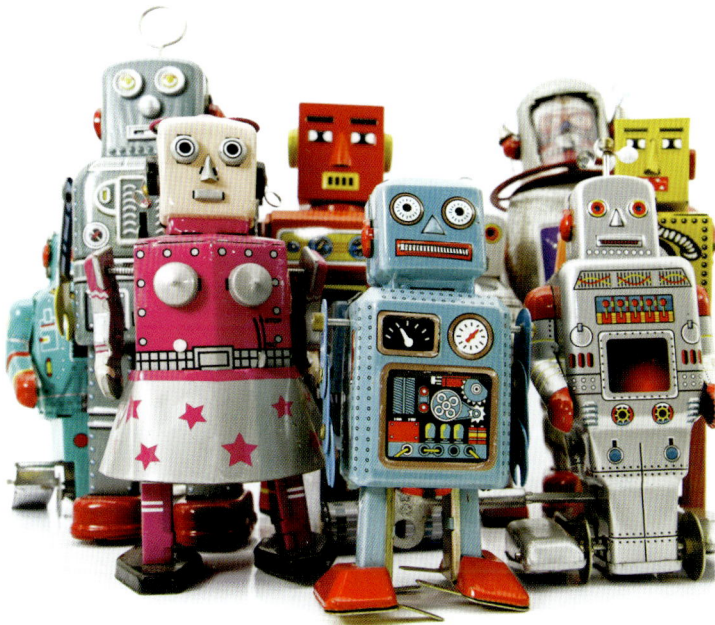

The power of robots working together

Two can do more than one, and cooperation pays off. Humans and animals cooperate in groups, herds, and flocks, and robots could benefit from doing the same. Working together enables robot swarms to achieve more in a shorter time: taking photographs inside a building that's about to collapse, for example, or cooperating to clean up a park. They are also less susceptible to damage: if one robot in a swarm is broken, the rest of the group can continue with the task.

Guido de Croon studies robot swarms at Delft University of Technology. He's inspired by groups of animals – mainly birds and insects – that display complex behaviour together. De Croon: 'By working intelligently together, ants can find food and bring it back to their nest via the shortest route. It is incredibly difficult for a single ant to know where the closest source of food is located, and to bring it back to the nest as quickly as possible. But hundreds or thousands of ants working together can find the shortest path.'

NATURAL FLIER: *The lightweight DelFly is inspired by a dragonfly and should be able to fly in a swarm.*
TU Delft

They don't do that by drawing a map of the surroundings, which might be the most obvious tactic for a robot builder. 'Ants leave behind a scent that fades over time, and other ants often follow that scent.' That way, a group of ants will eventually discover where food is located, along with the shortest route between the food and the nest; something an individual ant never could have accomplished. 'The fun thing is that an individual ant just does something extremely simple, but the effect at a larger scale is optimal behaviour. The ant automatically leaves a trail of scent, and if it has to make a choice about where to go, then it just follows its nose. So it's really something very simple. But when you put hundreds of ants together, they'll find the shortest path.'

A robot swarm without a boss

That's precisely the idea behind swarm robots: small, simple robots that together display intelligent behaviour. According to the strict definition, such a swarm doesn't have a boss or leader that tells its members what they should do, so there's really no way you can control a swarm. That makes it difficult to design, but extremely robust, because there is no single point of failure. De Croon: 'Say that you have a single leader, or a computer that controls everything. If that computer is deactivated or lost, then the whole system shuts down.'

You could consider self-driving cars in the flow of traffic to be a kind of swarm, even though the individual cars are far from simple. De Croon: 'For self-driving cars to function optimally, it would be more useful to have a single system that knows where they are all located.' Such a system might even help us never to have to sit in traffic again. 'But in practice, it's extremely difficult, and risky. If such a central system were to shut down, then you'd have a big problem.'

The same applies for GPS as well. The global system of navigation satellites can be attacked or hacked, and therefore disrupt the behaviour of the swarm. To prevent that from happening, a swarm of robots is better off simply not using it. De Croon: 'When you look for a solution, you notice that you end up very close to nature. That's not so strange, of course, because ants have done just fine without GPS for millions of years. A real swarm doesn't have any central system. Well, except of course the sun. Insects look at the sun in order to navigate, and when it's cloudy out, they use the polarisation of the sun's rays. You could say that what they see is a gigantic compass in the sky.'

De Croon strives to keep the individual robots as small and simple as possible. 'Animals have to use the energy at their disposal as efficiently as possible. The larger the brain, the more energy it uses. Energy efficiency is important. I think that's wonderful: they achieve an optimal general behaviour by following simple rules. I think that's very elegant. If you want to work with small, lightweight drones, you have to design their intelligence in very simple terms. So the question is: how can you solve extremely complex tasks using as little processing power as possible, with as few sensors as possible, and make them as lightweight as possible?'

Bees, for example, do all sorts of things that drones can't yet accomplish. De Croon: 'A bee can land perfectly on a flower, even when the wind is blowing, even though the wind is pushing both the bee and the flower in all directions. They can also find their way to a source of food and back home again over long distances, and they rarely hit obstacles in flight. Those are all still major challenges for drones.'

Yet bees are tiny, simple creatures, with only a million neurons — compared to the hundred billion in the human brain. De

Croon: 'If you were to ask engineers to build something that can do everything a bee can do, they would come up with something incredibly complex. But the bee probably does it completely differently. When biologists discover something about how those creatures work, it always turns out to be a very simple solution. I've always found that fascinating.'

SIGNALLING: *Swarm robots that communicate through light rings.*
Marco Dorigo/Université Libre de Bruxelles

Goal is mapping a building about to collapse

The RoboCup robot football tournament features several different competitions, varying from smaller robots that are all controlled by a central computer, to larger robots that each have their own brain. You could consider the latter to be a kind of small swarm:

the individual football robots decide what to do on their own, and they try to work as well as they can together, without a captain or coach telling them what they should do.

Real robot swarms are also flexible when it comes to size: you can add or subtract a few of the robots without changing the behaviour of the swarm as a whole. For example, when a swarm of robots is supposed to keep a park clean, it isn't a problem if one is hit by a car or attacked by a falcon once in a while. And on a busy day, you can add a few extra robots to take care of the extra rubbish.

In his research, De Croon usually tries to take on a challenging application. 'For example, the goal is to inspect a building that's about to collapse, so you don't want to send humans inside. That's difficult, because you have to work in an unexplored space.' Many researchers prepare such a space by placing beacons that emit a Bluetooth signal. Based on the direction and strength of the signals from a number of different Bluetooth beacons, the robots can calculate precisely where they're located in the building. 'But I don't want to do that', De Croon emphasises, 'because that's not realistic.'

In such a scenario, De Croon studies what the individual robots in a swarm would need to do in order to inspect the building together. The research involves a few steps. 'One: the robots have to be able to fly independently. Two: they have to be able to avoid obstacles. Three: they have to be able to avoid each other. We've been able to solve those problems fairly well so far. The next step is that they have to spread throughout the building on their own; a kind of basic swarm behaviour in which they fly away from one another. And then they have to come back, for example when their battery is almost empty. Once they're back, they can give us the photos of what they've observed inside the building.'

Another application is using drones inside greenhouses. De

Croon: 'Drones can explore where the fruit is ripe inside the greenhouse, or where water or pesticides are needed. It's a three-dimensional space: sometimes you want to look between rows of plants.' In that case, the fact that the drones are extremely light is a big advantage, because it makes them safe to use in environments where people are active as well, because they can't cause much damage if they accidentally fly into someone. A greenhouse is a much simpler environment than a collapsing building: you can easily install Bluetooth beacons or other adjustments before you let drones loose inside. And yet it is not so easy to make dozens or hundreds of tiny robots work together, even in such a protected, well-prepared environment.

Predicting how a robot will behave

So how can you build a robot swarm to perform complex tasks? It's already difficult to predict exactly what will happen when you bring together a large number of robots that behave a certain way. So many things happen simultaneously that it's likely that something unexpected will happen. It's also difficult to approach the problem from the other perspective: if you specify how a swarm should behave, what is the best way to program the individual robots? These are the two biggest puzzles in the field of swarm robotics at the moment: on the one hand, determining how an individual robot should behave based on swarm behaviour, and on the other, predicting how the swarm will behave based on individual robot behaviour.

De Croon explains that there are different tactics for creating a robot swarm. 'One way is to look at a natural system and try to imitate it based on what biologists know about it.' There are certainly advantages to imitating nature — if you copy a system that works, then you know that your system will work as well —

ROBOT DEVELOPMENT: *Two co-operating robots in a demo by Guszti Eiben at Discovery Festival in 2015.*
Bennie Mols

but does it actually make sense to literally apply principles from nature to artificial systems? Not necessarily, De Croon believes. 'Then you get drones that fly close together in a swarm, because birds can do it too, for example, but the birds might do that to avoid being eaten by predators. In that case, you'd be modelling unnecessary behaviour. On the other hand, we also ask robots to do things that aren't important in nature.'

De Croon therefore recommends focusing on self-learning systems. Ideally, you would give a swarm a common task, and then the individual robots could learn which simple rules to follow in order to display that behaviour. One way to do that is to allow the system to evolve, De Croon explains. In an example of such an evolving simulation, a swarm of a certain type of robots is put to work on a task. In subsequent evolutionary steps, the simulation would let the most successful robots 'reproduce' by combining their characteristics into a new generation of robots, which would improve the performance of the system. Eventually, you would arrive at a solution that works for the entire group, says De Croon.

That may sound simple, but it still leaves many questions unanswered. In a computer simulation, the evolutionary steps are faster and easier to make than when the researchers have to build new robots at every step along the way, but the simulations also have problems of their own. 'The problem with simulations is that they are doomed to succeed', De Croon adds. Simulations are easier than the real world. 'For example, in a simulation you can act like the robot knows exactly where its neighbours are located. But that's actually a very difficult step, even with small drones. We're still a long way from reaching that point.'

If swarm robots don't want to crash into their neighbours, then it is essential for them to know where those neighbours are. De Croon: 'Those drones are too small to map out their surroundings, let alone send their position on that map. Right now, we're looking at something resembling the scent signals that ants give off, but using technology.' The drones all emit a Bluetooth signal. The drone can then use the signal to estimate how far away another drone is, similar to the way you can see how far your telephone is from a wireless antenna or WiFi router based on the number of bars on the screen. 'A drone knows how fast it's flying and in what direction, and it sends that information to other drones. The others can then use that message, plus the strength of the signal, to know approximately how far away they are.'

De Croon's research group was the first in the world to succeed in making swarm robots measure the distance between them without using separate Bluetooth beacons. The technology doesn't provide precise distances, but it works well enough to help tiny drones weighing around 40 grams to avoid one another in flight. 'Now we know that the drones don't know exactly where the other drones are located. The great thing is that we can use that insight to create even more realistic computer simulations.'

Newer sensors and technologies may also provide a boost to the field of swarm robotics. Researchers in Switzerland are currently working on robots that can communicate with one another via ultra-wideband, a new technology that makes it possible to transmit large amounts of data quickly over short distances, and that also uses little energy. That makes it ideal for swarm robots. Using the technology, the researchers were able to have swarm robots determine their distance to one another with much greater accuracy. De Croon: 'They did use beacons in their experiments, however. We're now applying those developments with ultra-wideband to our method for determining relative position without those beacons, to make it more accurate.'

LEARNING TO SWARM: *Two robots from Zebro Project TU Delft, in a demo at Robo Business Europe in 2017.*
Bennie Mols

174

Robot swarms in the real world

For the moment, robot swarms mainly exist inside computer simulations and robot laboratories. De Croon therefore tempers his enthusiasm about the utility of robot swarms. 'Actually, we don't really know for which situations swarms would be the best solution. There are clear advantages to the principle of a group of robots in which losing one isn't a big problem, but it's not yet entirely clear what the 'killer app' will be.'

Nevertheless, roboticists are already thinking about what will eventually be needed to use robot swarms in the real world. Predictability, for example: that the robot swarm will do exactly what it is supposed to do. The research group led by Radhika Nagpal, Professor at Harvard University, has developed a way to prove that a swarm will display a specific behaviour based on the behaviour of its individual swarm robots. De Croon: 'That's very interesting for real-world applications. It might be acceptable to say: we've tried it a thousand times, and it's worked every time. But when you actually put a system to use, it's a bit more comforting if you can mathematically prove that it works.'

Nagpal has developed several robot swarms, including Kilobots: a swarm of 1024 robots. The robots communicate with one another via infrared signals. They have three metal legs that they use to walk and recharge via the floor. When you give one of the robots a goal, such as building a specific structure, it will pass the message on to the rest like a game of Chinese whispers. In another project, Nagpal developed termite robots that could build a structure together without human supervision. The robots in the project were given 'traffic rules' by a central computer, but they then decided for themselves where to put different blocks.

Nagpal was able to prove that her swarm did what it was supposed to do: the robots built a predetermined structure based on their individual programming code. De Croon: 'Strictly speaking, they weren't real swarms, because the robots were all told which form they were supposed to build. On the other hand, from a practical application point of view, it is good to be able to say: now I want them to build a castle.'

That's exactly the idea that Nagpal had in mind. With a swarm of tiny robots, you might be able to produce smart, programmable materials and structures, a bit like a dynamic 3D printer. And perhaps in the future, our houses might be made out of multifaceted robots, so you could add a wall when a guest comes to stay, or move a balcony to follow the sun around the house, at the press of a button.

A robotic swarm looks for a queen

Robot swarms with a leader are not considered true swarms in the strictest sense, but some natural swarms do have them: every beehive has its queen, after all. The Borg, the cyborg community in *Star Trek*, also has a Borg Queen as its leader. The Borg travel through space searching for new beings to assimilate in order to make their community stronger and more intelligent. All of the new skills added to the Borg are shared with all members of the community. The Borg can therefore do quite a bit, but in exchange they lose their free will and are controlled by the Borg Queen.

That's how many swarms work, and the benefits are clear: the members of a community can use the knowledge and skills of their colleagues, and with a Great Leader as the head of the community, you don't have to deal with the limitations of the individual.

Amazon uses such a cooperative robot system to move goods around its warehouses. In the company's warehouses, tens of thousands of robots that look like bright orange robot vacuum cleaners drive under racks of products, lifting the entire rack in order to bring it to another location. A central computer monitors the locations of all of the robots and goods in the warehouse.

When the central computer receives the command to pick a specific item, it passes the command on to the nearest free robot, along with the location where it can find the item. The robot uses QR codes on the floor to navigate to the item, and then brings it to the delivery point to have it packaged in the same way. Sensors keep the robots from bumping into one another. Along the way, they pass on their observations to the central computer, which collects information about things like dips in the floor or crooked stickers. The system appears to work well: the company has reduced the time between the order and the shipment from one hour to 15 minutes.

But every robot in the crowd is unique

Would a swarm of different types of robots be more effective than a swarm of robots that all do the same thing? In the Swarmanoid project, researchers studied heterogeneous swarms, which consist of robots that can perform different tasks. The project combines the efforts of 'hand-bots', 'foot-bots', and 'eye-bots'. When the system receives the command to pick up a specific book, eye-bots fly to the ceiling to see where the book is located, after which a few mobile foot-bots trace a route to the bookshelf. Two other foot-bots then drag a hand-bot along this route to the bookshelf, and the hand bot climbs the shelf to take out the book. It's a fairly complicated collaboration if you can't program who should perform each task in advance. How can you develop such a system? De Croon mainly sees possibilities for self-learning systems: 'The best thing would be able to say: this is the task, these are the robot components that we have, and then let the system design the rest on its own.'

De Croon explains that in real life, swarms never actually consist of fully identical robots. 'If you have a hundred drones, then one of them will have a slightly different sensor, another will have a broken compass... When you look closely, it is actually a heterogeneous swarm. They're all just slightly different from one another. The same applies to humans and animals as well. Ants are also all just a bit different from each other. No swarm consists of truly identical individuals.' Is that a bad thing? No, on the contrary. 'When you cooperate, you can actually make good use of different characteristics. They can help you go a lot farther.'

Collaborating with your future workmate

The teacher of the future will not only have to be able to deal with children, but also with robots, predicts Serge de Beer. He began his career as a secondary school engineering teacher, but today he mainly focuses on using robotics in education. He teaches courses on the subject. De Beer: 'The funny thing is that many of my students imagine a highly multifunctional robot that can mark tests during the day as well as clean up the canteen after hours. They're very aware of the fact that a robot can work all day and all night.'

Is that how De Beer sees the future as well? 'The longer I work with robotics, the more I'm convinced that robots will have an influence on education. Initially as a subject. I wouldn't be surprised to see subjects like robotics or artificial intelligence taught at school.'

His fascination for robots began at a young age, when he visited an event organised by the famous Dutch inventor and television presenter Chriet Titulaer. But his interest died down after a while. 'For decades, only the field of industrial robotics made any progress. Over the past few years, the media started talking about intelligent robots again, and my old fascination came back. It has always surprised me that it took so long.'

Instead of studying robotics, De Beer chose to become an engineering teacher. 'I love education. And every facet of technology is represented in robots: the mechanical, the electronic, and the ICT. Robots also perform a lot of technical work. Today, I teach humans to use technology, but in the future I'll be teaching that to robots too.' De Beer sees a role for himself in two areas of education in the future. 'On the one hand, I want to help people use robots in education. But I also want to train and test robots as well. At the moment, I'm working on a robot toolkit that you can use to create a personal development programme for a robot, for example.'

De Beer predicts that in the future his students will work together with robots. 'I don't believe in fully autonomous systems.

Instead, I think we'll have pre-programmed missions, with humans as a kind of robot manager. In the future, a plumber might have a robot assistant, for example. I think that all of today's students will eventually work with a robot.'

The same applies to teachers, De Beer believes. 'I see robots more as teaching assistants than as fully-fledged teachers. You can see that today with the different types of Nao robots, which can already help students practice their vocabulary.' De Beer is currently developing a link between digital education systems and robots. 'That way, the robot will know that this particular student has difficulty with fractions, for example, as well as how far along in the material he or she is. Then the robot wouldn't be just a gadget anymore, it would be a true colleague.'

But De Beer thinks that there are some areas that need improvement before that point is reached. 'The main barrier is still the sophisticated communication between humans, such as ironic comments. But a robot should eventually be able to do the physical work that we humans can do just fine. When I clear out the dishwasher, and I pick up a cup that's stacked awkwardly under another cup, I think: how would a robot solve this problem? For a robot, that would be an extremely complicated task.' And once we've solved those kinds of problems? 'Robots will eventually be able to do much more physical work than humans can. A robot can keep going day and night. It can also become proficient in a language automatically by downloading the right file.'

But we're not there yet, says De Beer. 'During my workshops, I always show the DARPA fail video to reassure people who are afraid of losing their jobs to a robot.' The video is a compilation of the biggest mistakes during the DARPA Robotics Challenge contest. 'For example, you see a robot that has to turn off a fire hydrant. The robot stands next to the hydrant, makes a turning motion in the air, then falls over. No, I don't think we'll have anything to worry about for the next 10 years at least.'

11 The importance of building ethical robots

Isaac Asimov's three rules about rogue robots

In the film *2001: A Space Odyssey*, the protagonist Dave is locked out of the spaceship taking him to Jupiter by the intelligent on-board computer HAL 9000. HAL feels threatened by Dave and his crewmate, and uses that as a reason to try to eliminate Dave. As HAL explains: 'This mission is too important for me to allow you to jeopardize it.' In short: HAL sees itself as indispensable for the mission to Jupiter, and the mission is more important than the lives of its human passengers.

HAL 9000's unethical behaviour proved to be a source of inspiration for countless films and stories featuring computers or robots that pursue their goals without any regard for the humans who cross their paths. But this dystopian image of our future with robots and artificial intelligence would never come to pass if robots abided by the Three Laws of Robotics devised by science fiction author Isaac Asimov in 1942:

1. A robot may not injure a human being or, through inaction, allow a human being to come to harm.

2. A robot must obey the orders given it by human beings except where such orders would conflict with the First Law.

3. A robot must protect its own existence as long as such protection does not conflict with the First or Second Laws

Robot stories written before the 1940s often featured a plot like that of Frankenstein: a robot that's initially friendly is provoked at a certain point and rises up against its maker. Asimov thought that was nonsense. He saw it as an expression of a technophobic world view that he could not accept. In Runaround, one of the stories in the anthology *I, Robot*, he described the laws of robotics for the first time.

But later in life, he was relatively indifferent about his robot laws. 'The Three Laws were actually obvious from the start', he wrote in Compute! Magazine in 1981, 'and everyone is aware of them subliminally. The Laws just never happened to be put into brief sentences until I managed to do the job. The Laws apply, as a matter of course, to every tool that human beings use.' After all, a hammer also isn't intended to injure humans, and is expected to do only what you want it to do.

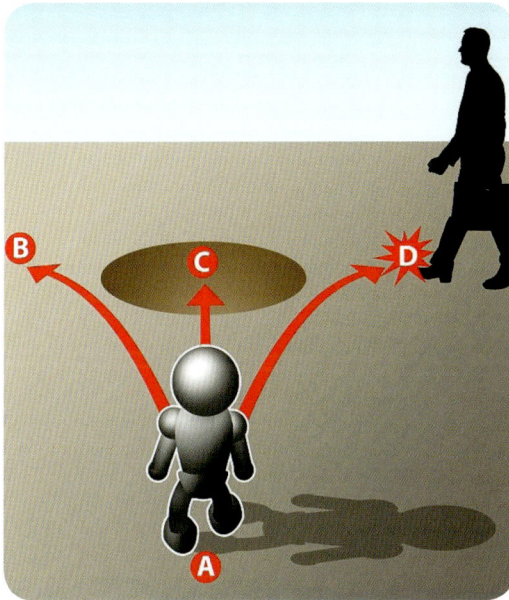

ETHICAL ROBOT: *Following Isaac Asimov's first rule, this robot must do everything in its power to stop someone from falling into the hole.*
Alan Winfield

Asimov later added a 'Zeroth Law', which was more important than the first three:

Zeroth Law: A robot may not harm humanity, or, by inaction, allow humanity to come to harm.

Imagine that a robot receives the order to guard the launch button for a battery of nuclear missiles. What should the robot do if a malicious intruder tries to reach the button, and the robot's only options are to eliminate the intruder or allow him to press the button? According to Asimov's original three laws, the robot would not be allowed to injury the intruder — who is a human, after all — because that is more important than following orders... Kaboom.

The Zeroth Law would prevent such a disaster scenario from occurring. Asimov admitted that it is difficult to define what 'harming humanity' entails, precisely. Yet he still included the Zeroth Law in two of his books.

When robots go wrong

Say that a robot misbehaves anyway, and breaks the laws of robotics? Isn't it just programmed wrong? Doesn't the responsibility lie with the robot's maker? Not if the robot can decide who to injure or not to injure, says researcher and artist Alexander Reben. All forms of 'dumb' technology have a human at the controls, and it is therefore clear who is at fault if a person is injured, he says, but the issue is less clear-cut if a robot can act on its own accord.

So in 2016, Reben built a robot that can decide independently who to harm. A robot arm equipped with a needle sits on a box the size of a large lunchbox. If you hold your finger in front of the robot, it will decide — purely at random — whether to prick your finger or not. Reben named the robot First Law, after Asimov's first law of robotics. He built the robot as a philosophical experiment in order to spark a discussion about robots that can decide independently whether or not to harm humans. 'The robot makes a decision that I, as the maker, cannot predict', Reben explains. 'I don't know if it will hurt you or not.' That makes the robot 'ethically dubious'.

In order to prevent autonomous robots from acting maliciously, Reben thinks that it may be necessary to install a kill switch on every intelligent machine. But that presents you with the problem of HAL 9000 in *2001: A Space Odyssey*: if the robot is so smart that it is aware that it can be shut off, will it not do everything in its power to prevent that? Asimov's laws of robotics are intended to prevent exactly that kind of behaviour.

But it's not that simple: in order for a robot to obey Asimov's laws of robotics, the robot must at least be aware of the consequences of its actions. To do that, the robot would have to be able to predict the future, says British researcher Alan Winfield from the Bristol Robotics Lab. In 2014, he built a robot brain with a 'consequence engine'. Before the robot decides which action to perform, it forms a kind of mental image of all of the possible actions and the consequences that those actions will have, both for itself and for the rest of the world, including other robots and humans. The consequence engine does not decide which action the robot should perform, but it does state which actions will have undesirable consequences, so the robot will at least avoid those actions.

With this consequence tool, Winfield has created an 'ethical robot'. The robot drives around together with other robots on a field scattered with pits. The ethical robot follows its own plan until one of the other robots threatens to fall into a pit. In that case, the ethical robot will change its course to give the other robot a push in order to change its course away from the direction of the hole. Winfield has even programmed the robot to help humans before helping another robot.

But when it has to choose between two robots that are both headed towards a hole, in more than half of the experiments conducted it spent so long thinking about which one to help that it was unable to save either robot. Winfield thought that he had built an ethical robot, but in practice the robot didn't always display the most ethical behaviour. After all, it would be better to save at least one of the two robots than to let both of them fall.

That touches on the 'trolley problem', which is often referred to as a major ethical problem in the context of self-driving cars. If a car is headed towards a bus full of children, and can only avoid a fatal collision by driving into a ravine and killing its own passenger, what should the car do?

Roboticist Rodney Brooks isn't impressed with this so-called 'dilemma'. 'How many times when you have been driving have you had to make a forced decision on which group of people to drive into and kill? You know, the five nuns or the single child? Or the 10 robbers or the single little old lady? For every time that you have faced such decision, do you feel you made the right decision in the heat of the moment? Oh, you have never had to make that decision yourself? What about all your friends and relatives? Surely they have faced this issue? And that's my point. This is a made up question that will have no practical impact on any automobile or person for the forseeable future.'

Responsible roboticists are planning for the future

Asimov's laws are based on the assumption that robots will behave as if they are humans, but we're nowhere near that point yet.

Robots are designed and built by humans, so a robot that behaves ethically begins with ethical robot designers and builders. They are the ones responsible for a robot that not only does what it's told, but that also does so safely and in accordance with the norms and values that we as a society consider important. Roboticists have to take those norms and values into consideration from the very start of the design phase.

In 2010, a group of British scientists from the fields of robotics, ethics, law, social sciences, and the arts came together to develop a practical alternative to Asimov's three laws. Rather than laws for robots, the group came up with five ethical principles for the humans who design, build, sell, and use robots. These principles are not etched in stone; the scientists see it more as a living document that can be refined based on real-world experience.

According to the first ethical principle, robots are multifunctional instruments that may not be designed purely to kill or harm humans, except in the interest of national security. Instruments can often be used for multiple purposes. A bread knife, for example, can be used to slice bread, but also to murder someone. In the same way, a drone that was designed to deliver parcels can also be used to drop an explosive device over a crowd of people. But according to the first principle, drones should not be designed specifically for that purpose, with the exception of military drones that serve the higher purpose of national security.

Secondly, robots should be designed and used in accordance with existing legislation, fundamental human rights and freedoms, and with respect for the user's privacy. With regard to privacy: robots collect data about what they observe, but what happens with that data? A care robot can collect data about patients 24 hours per day. That can come in handy, but how long should the data be stored, and who should be allowed to access it? A toy robot can conduct conversations with a child, but what if the manufacturer records those conversations? The moral responsibility for what happens with that data lies with the human, and not with the robot.

According to the third principle, we should see robots as products that must meet requirements for safety and reliability, just like any other consumer product. Robots embody information technology, so they're also vulnerable to hackers. The third principle therefore also implies that robots should be protected against attacks by hackers.

Fourth, it should be clear to the users that a robot is an artificially produced product. Robots should not be designed to deceive vulnerable users, such as children or lonely elderly people, to believe that the robots are real people or animals. Humans are so good at anthropomorphising what they see around them that robots can easily give the idea that they have real feelings, emotions, and intentions that are identical to human feelings, emotions, and intentions. Naturally, users may enjoy experiencing such feelings, and it may be a useful purpose for robots such as care robots, but according to the fourth principle, users should be able to understand that they're dealing with machines, and not with living beings. Imagine that an elderly person suffering from dementia can't tell the difference between a care robot and a real person; according to this principle, using the robot would be unethical.

The fifth and final principle stipulates that one or more people must be held legally responsible for the robot's behaviour. A robot should have some kind of identifying mark, like a car's registration plate. If the robot doesn't do what it was designed to do, or if it causes an injury, then a human must bear legal liability for the damages, not the robot. Although robots will become increasingly autonomous in the future, most roboticists believe that they should always be supervised by humans.

The five ethical principles that roboticists should abide by can be summarised in the sentence: humanity should flourish more with robots than it would without them.

Robots and the UN's development goals

If we treat robots responsibly, they have a lot to offer us. In 2015, the UN formulated 17 sustainable development goals for the period 2015-2030. The list includes eliminating poverty and hunger, providing good health and education for all, and equal treatment regardless of gender, ethnicity, and socio-economic status.

Robotics can make real contributions to realising several of these goals. The UN projects that the world population will increase from 7.5 billion people in 2017 to 10 billion by 2050. Those additional 2.5 billion people will all have to be fed, but the Earth will not get any larger. Robots can help feed all of those extra mouths in a number of ways. For example, drones, in combination with satellites, can help predict harvest failures. Drones and other agricultural robots can also contribute to precision agriculture, in which each individual animal and plant receives the exact care, nutrition, pesticides, and medication it needs at a specific moment in time. That will save energy, water, and pesticides, while increasing agricultural yields. Self-driving cars and drones can also play a role in transporting food to hard-to-reach areas.

Even today, many people in the world have to perform demanding or dangerous physical labour. Think of the building industry, mining, or cleaning industrial complexes. Robots are ideal for helping people perform these tasks, or may even take them over entirely. Some roboticists even consider it a moral duty to have robots take such work off of our hands.

One billion people on earth have some form of physical handicap. Robotic aids, such as exoskeletons, can help these people regain their freedom of movement.

In the field of sustainability, robotics can contribute to the creation of a circular economy, in which materials are re-used intelligently. Robots can also be used to protect and preserve nature on land, at sea, and in the air. The wide-scale introduction of self-driving cars and lorries is expected to cut energy use for road transport by 20 per cent and double road capacity.

The 17 UN development goals are admirable ideals, but the world's citizens will only accept robots at work and in their daily lives if they can trust the robots, as well as the organisations that regulate their design and construction.

How will robots change the human race?

So far, we've examined how robots and their makers can and should behave ethically. But what about the users of robots? Should we be allowed to do anything we want to robots? In the television series *Real Humans*, people live in general harmony together with realistic android robots that serve as assistants, household help, babysitters, or co-workers. The differences between humans and robots are minor, but they are exaggerated by the way they deal with each other. Some people treat robots with casual disdain, exemplified by the man who thoughtlessly squeezes his robot secretary's breast in passing. 'Almost real', he says tauntingly. The robot in question seems unfazed by the harassment.

Is that a bad thing? Or to put it more precisely: if a robot doesn't feel emotions, what does it matter if you mistreat it? It might not be nice to kick your bicycle in frustration when it has a flat tyre, but it isn't unethical. But if you were to treat your pet the same way, then it would be undeniably so – and even punishable by law.

Imagine how it feels when you accidentally drag your old-fashioned vacuum cleaner down a step. Now imagine how you would feel if you saw your robot vacuum cleaner fall down the same step. In both cases, you would probably mainly be concerned about whether or not the appliance is damaged. You wouldn't feel sorry for your old-fashioned vacuum cleaner after such a fall, and you probably wouldn't feel sorry for the robot vacuum cleaner either. After all, the robot doesn't have emotions or consciousness; it's more of a household appliance than a living being. So you treat it as if it were an inanimate object.

The boundaries fade, however, as a robot begins to resemble a human or an animal, even if there are no indications that it has emotions or a consciousness. When you see someone kick a robot dog, it feels wrong, even if you can clearly see that the robot is made from mechanical components. Is it ethically wrong to kick a robot dog if you are absolutely certain that it is not conscious and cannot feel pain? Probably not. Yet it still makes us feel uncomfortable. And that's the key: the way we treat robots also has an effect on us.

A large-scale meta-study by American researchers has indicated that humans who play violent computer games behave more aggressively. Test subjects who had just played a violent game appear to be less prone to help other people, and are quicker to make word associations with aggressive subjects. Some studies show that this effect fades after a time, but even if the effect is short-lived, it is absolutely clear that technology influences how we think and how we behave.

If we treat robots that look like humans or animals as if they were lifeless objects, it may also have an influence on our own social inhibitions. When you are accustomed to kick your robot dog, or squeeze your robot secretary's breasts, then you unconsciously learn that kind of behaviour is acceptable, and you will start to treat real animals and humans differently.

Technology has even penetrated into our intimate lives — from the introduction of the first vibrators at the end of the 19th century, to 'teledildonics': sex toys that you can operate via the Internet. It seems as if humanity isn't afraid to take technology to bed. It's therefore entirely possible that the same may happen with robots in the future as well. In the 17th century, seamen were already making dolls out of cloth to escape the loneliness of long sea voyages, and recently realistic 'companionship dolls' have been developed that can make sexual movements.

A robot in bed; isn't that a bit frightening? It could be an option for people who can't easily look for a partner themselves, such as the elderly or people with a serious physical or mental disability. And perhaps prostitution in its various forms could be replaced by robots as well. On the other hand, sex robots could make their users feel even more lonely.

Psychologist Sherry Turkle at Massachusetts Institute of Technology has concluded that technology changes us. We increasingly contact one another via technology; instead of making arrangements to actually meet, we send text messages or emails. The next step, Turkle says, is to replace the other person with technology entirely. It's now so easy for us to interact with technology that an artificial companionship robot is no longer such a strange idea. At a TED presentation in 2012, she emphasized our intimate relationship with technology. 'We want to spend time with machines that seem to care about us. We're designing technologies that will give us the illusion of companionship, without the demands of friendship.' The robot might disappoint us, but we will never disappoint the robot.

Is that a bad thing? In her book *Alone Together*, Turkle discusses the example of a robot seal used as a companion animal for elderly people. At first glance, it seems like a good solution for loneliness. The robot seal is perhaps better than nothing, but

Turkle's research shows that elderly people would always rather have a human across from them than a robot. The elderly people who participate in research on companion robots are mainly pleased by the attention they receive from the researchers and the feeling that they are contributing to something.

Not everyone agrees with Turkle. In a video on his own YouTube channel, Alexander Reben, the creator of the First Law robot, says that it is actually very normal to have non-human buddies. 'A lot of people look at things like robots and say: well gee, this social robot's going to take away my relationship with people. But we actually can point back to some examples like the dog, which was bred from the wolf to be a companion for humans. I think there's a lot of history we can look back on and actually see that this is not a very unnatural thing for us to do.'

Keeping pets is commonplace, and for most people — with the possible exception of 'cat ladies' — they aren't a replacement for humans. That may be how we come to see robots in the future too: not as full replacements for humans, but just as another creature, with its own pros and cons. The companion robot may eventually become a kind of pet, which we treat humanely and with love, but which isn't one of our fellow humans.

Killing machines: robots in the military

Robotics is also on the rise in the field of military technology. When the Soviet Union launched the satellite Sputnik in 1957, the United States was caught completely off-guard. Never again, thought the American President Dwight Eisenhower. To that end, a year later he founded the Advanced Research Projects Agency, or ARPA for short. In 1972, the agency was re-named DARPA: Defense Advanced Research Projects Agency. Since then, the

WORKING FOR THE ARMY: *Robots have many different military uses, including defusing bombs and searching for people trapped in buildings. This is a PackBot.*
iRobot

military research organisation has been one of the driving forces behind the development of new military technology, including military robots. DARPA has an annual budget of $2.8bn, making the US by far the largest player in the field of military robotics.

In the early 21st century, DARPA also organised contests to develop self-driving cars (the DARPA Grand Challenge) and to develop humanoid robots for disaster zones (the DARPA Robotics Challenge). Naturally, the US also hopes to be able to use these civilian robot technologies for military applications.

Some of these applications deal with unarmed robots, such as those that can disarm improvised explosive devices, but armed robots are also under development. These are often colloquially referred to as killer robots, or 'killbots', but the military prefers the term Lethal Autonomous Weapon Systems. Whatever one calls them, the use of armed robots raises countless ethical questions. How much autonomy should military robots have to kill humans? How are they subject to humanitarian law of war? Will they lead to a new arms race for increasingly intelligent weapons?

In 2005, the American Department of Defense developed a long-term vision in which the military will increasingly use more and more autonomous weapons. At the moment, the only ones in use are remote-controlled systems, such as drones, but gradually military systems will be able to make more decisions on their own. The department expects to deploy the first fully autonomous weapon systems by the year 2050. These systems will select their own targets within a pre-designated area. Once they have been activated, humans will no longer be able to control them.

The development of robotic weapons has progressed more rapidly than people had expected around the year 2000. In 2002, the Americans conducted the first attack with a drone in the fight against the Taliban in Afghanistan. So far, the drones have all been controlled from inside the United States, but an aircraft that can identify and destroy enemy targets on its own is already under development: the X-47B. A prototype has already flown, and a more advanced version may be airworthy by as early as 2020. Other countries, such as Russia, Israel, China, India, France, and the United Kingdom, are also developing their own fully autonomous military robots.

South Korea already has a kind of robot sentry deployed along its border with North Korea (the SGR-A1). These are basically static machine gun positions equipped with heat sensors and movement detectors. They are capable of identifying and firing at intruders within a radius of 500 meters. At the moment, they are still under human control, but from a technical perspective it's already an easy matter to remove the human from the decision-making loop entirely.

The Israeli army is equipped with the Guardium robotic armoured vehicle, which patrols the border with the Gaza strip and can fire autonomously. This system is also operated under human supervision at the moment (and is subject to international pressure), but in principle, the system also allows humans to leave the decision to pull the trigger to the robot.

Supporters of autonomous weapons have presented a number of arguments in their favour. For example, autonomous weapons can save lives on their side in a conflict, because they are unmanned. But they can also save lives on the enemy's side, because they can conduct more precise attacks. The idea is that the robotic weapon's many sensors and its ability to collect several different types of data enable it to make more rational and ethical decisions, because they can see through the traditional 'fog of war'.

Autonomous weapons could also help cut costs, react quicker, never get tired, never act in vengeance, panic, or anger, act as a deterrent against states that don't have similar weapons, and be used in areas that are inaccessible to humans.

Opponents of autonomous weapons counter with the moral argument that only humans should be allowed to decide on matters of life and death, and not robots. Robots can't understand the context of a conflict, have no moral understanding of human values, and don't comprehend the human motivations behind

MORE HUMANE? *Some believe that drones could save lives by targeting bombs, but others are alarmed by the rise of remote-controlled warfare.*
Wikipedia Public Domain

their deeds. Another question is if robots can assess whether an opponent is seriously wounded or about to surrender. That's important, because the humanitarian laws of war prohibit shooting at soldiers who are wounded or trying to surrender.

The laws of war also stipulate that the use of violence must be in proportion to the expected military advantage. It's doubtful that such a consideration could be left to a robot.

Autonomously operating weapons could also lower the threshold for the use of violence, and encourage greater risks, such as flying lower than human pilots would be willing to fly. And since they can hit their targets with greater precision, more targets could be hit, which may in turn lead to more innocent civilian casualties.

That raises the question of how the local population would react. If unmanned robotic weapons are seen as 'cowardly', then it may reduce support for ending the conflict, or the country that uses autonomous weapons may become the target for terrorist attacks.

In the distant future, autonomously operating weapons may also have the capacity to learn, which would make it even more difficult to answer the question of who is responsible for the weapon's decisions.

Since 2013, the international organisation Campaign to Stop Killer Robots has advocated for a prohibition on the use of autonomous weapons, similar to the existing prohibitions on the use of biological and chemical weapons, or specific technologies, such as laser weapons intended to blind opponents or land mines for use against personnel, and not equipment such as tanks. Co-founder and emeritus Professor of Robotics Noel Sharkey says that the list of possible errors in the use of autonomous weapons is far too long to ignore, and that there's a limit that humans should never cross: 'Robots should never have the authority to kill humans.'

In 2015, the Future of Life Institute started a petition against the use of autonomous weapons, with support from academics specialising in artificial intelligence and robotics. The petition, signed by a large number of leading scientists, concludes with the words: 'We believe that AI has great potential to benefit humanity in many ways, and that the goal of the field should be to do so. Starting a military AI arms race is a bad idea, and should be prevented by a ban on offensive autonomous weapons beyond meaningful human control.'

The social impact of robotisation

Catelijne Muller is a member of the European Economic and Social Committee (EESC). For many years, Muller has used her legal background to advocate on behalf of workers' rights. In the spring of 2017, she issued an advisory report on artificial intelligence, based on interviews with experts.

Muller: 'Digitisation has been on the agenda for quite some time, and I occasionally read articles about artificial intelligence. When I read that people like Stephen Hawking and Elon Musk see artificial intelligence as 'a threat to humanity', I decided to learn more about the subject. In order to reap all of the benefits it has to offer, we will also have to understand the risks and challenges it presents. Otherwise, we'll start to get all sorts of regulations written out of panic and a lack of knowledge, and nobody wants that to happen. So I try to prevent it by getting ahead of the game.'

How should we react to the concerns that many people have about robotisation and artificial intelligence? Muller: 'People often feel so overwhelmed by technology, but that's entirely unnecessary. We as a society have a choice. People are too quick to say: robotisation is great, because robots take over the boring, dirty, or dangerous work, so that we have more time to do jobs we enjoy more. That's fine, but what happens if an employer thinks: great, now I can lay off all of those employees. Employers should assume their responsibility and not shove those people aside, but rather use them to expand the company and develop new things. That's good for both the company and for society as a whole. On the other hand, employees will have to be flexible and accept a certain degree of robotisation.'

Muller's report was well-received. 'The European Commission thought it was 'down to earth' – but not in an insulting way. I definitely wasn't insulted. Perhaps they were worried it would be too alarmist, but it's not that at all. For example, we shouldn't focus too much on super-intelligence. There are so many things that can be done right now.'

Muller therefore got to work immediately. 'We should strive to formulate standards for artificial intelligence, just like we do for food, washing machines, and other consumer products. The standards should deal with the system's security, but also with verification, responsibility, and respecting our ethical values and fundamental human rights, such as privacy. I'm the first person at the European policy level to ask for those kinds of standards. It takes a while to develop standards and get them accepted.'

But can artificial intelligence and robots be compared to washing machines? Muller: 'The thing that's fundamentally new about artificial intelligence is the self-learning aspect. We can never completely understand why a self-learning system displays a specific behaviour.' It's therefore vital that we start thinking about who should bear responsibility and liability for a self-learning system.

Muller: 'A few months before my report was published, the European Parliament issued its own report on artificial intelligence. It's similar to my report in most areas, except that they want to commission a study on whether you can make intelligent robots a legal entity. The idea behind that is that the maker of a self-learning system no longer controls what the system teaches itself over time, and should therefore not be held liable for it.'

Muller strongly disagrees. 'You should never go down that road. Liability also has a preventive effect: developers should feel obliged to bring something good to the market. You shouldn't be able to say: this is an intelligent system, and it bears its own responsibility. If you do, then manufacturers will grasp any opportunity to say: 'But it's got artificial intelligence!' in order to avoid liability. I always compare it to vicious dogs. The owner is responsible for them.'

12 2100 — A Robot Odyssey

The future of work in a robotic world

The International Federation of Robotics (IFR) predicts that sales of robots will increase by between 10 and 20 percent per year for the foreseeable future. Their prognosis includes both industrial robots and service robots purchased by consumers and businesses. What effect will this rapid growth in robotisation have on the work that people will do in the future?

One study that has received considerable media attention over the past few years, is the report 'The Future of Employment' published by Oxford University researchers Carl Benedikt Frey and Michael Osborne in 2013. They calculated that 47 percent of all of the jobs in the US could be automated within the next 20 years. Middle-class jobs would be especially hard-hit. Most of the automation would involve computers, and only a small proportion of the jobs would be taken over by robots, as they can only perform predictable physical tasks at the moment. In 2014, Deloitte conducted a similar study for the Dutch economy, and reached similar conclusions.

The worrisome results of the two studies should be taken with a grain of salt, says Robert Went from the Netherlands Scientific Council for Government Policy (WRR). Went was the lead author of the WRR study 'Mastering the Robot', from 2015. Went: 'Frey and Osborne defined 'jobs' in a far too simple manner, as if a job consists of only a single task. In reality, the vast majority of jobs involve bundles of tasks. I would emphasise that point even more today than we did in 2015. Some of those tasks can be automated, while others simply cannot. You can often automate how you search for information, but it's impossible to automate social interactions, such as giving a presentation or coordinating with colleagues.'

Since the results of the Oxford study were published, groups such as the Organisation for Economic Cooperation and Development (OECD) and the consultancy bureau McKinsey & Company have conducted new studies that consider jobs as bundles of tasks. These studies have produced results that Went believes are far more realistic: 'According to the OECD, over the next 20 years, no more than nine percent of all jobs will be fully automated. McKinsey puts it at less than five percent.'

Although only a limited number of jobs will disappear entirely, the WRR study concludes that the nature of the work will change for around 60 percent of the jobs. Went: 'For example, a judge will be able to search accurately and reliably through case histories. That will give her more time to prepare the verdict. From an economic perspective, it would be best not to use robotisation to cut labour costs, but to have the employees use their newfound free time to create new added value. They can call clients to offer more personal service, for example. Or think about how to improve a production process, which is something a robot can't do.'

The economist emphasises that robotisation is not a natural phenomenon that we have no control over. 'Governments, businesses, social partners, but also engineers can steer robotisation in the desired direction. Take robots in health care or education, for example. We don't have to strive towards robots replacing humans. We can deliberately invest in robots that supplement teachers or caregivers. That's both desirable and more realistic.'

Moreover, every trend will generally provoke a counter-trend. On the one hand, supermarkets are increasingly introducing self-checkout services, which makes till staff redundant. On the other, the Apple Store has plenty of well-educated experts on the sales floor, who offer customers fast, informed, and personal service.

Went: 'Predictions about the rise of robots are often wildly optimistic. I've learned to divide the enthusiastic predictions by around 10, and they usually come closer to reality. Take robot pizza delivery vehicles on university campuses. Great idea, but they're all still very small-scale applications. Or look at the experiments in China with robots that take orders and carry food and drinks. The robots were quickly 'laid off', because they would bump into customers, spill their orders, and had all sorts of other problems. On the other hand, some McDonald's locations are experimenting with offering table service. Many people still appreciate that, and some elderly customers don't want to have to carry a serving tray.'

Based on these conclusions on how work will change in the future, WRR has issued four recommendations. 'First, we have to invest in robotics', says Went. 'As the population of many Western countries begins to age, we'll start to need robots more and more. We'll have to produce more goods with fewer workers, and we'll have to care for an increasing number of people who can no longer work.'

When investing in robotics, Went believes it's crucial that robots are developed in collaboration with the people who will be using them in practice. Ask workers what they need, and don't just build something because it's technically possible. He gives the example of a device that Philips had developed to remind senior citizens to take their pills. It worked fine from a technical point of view, but the seniors didn't like to use it because it reminded them that they are alone every day, instead of having a family member or caregiver remind them to take their pills. Went: 'After that, Philips realised that they perhaps needed fewer technicians and more psychologists and sociologists.'

Second, robotisation will mean that many people will need re-training. People will have to learn how to work together with robots. They will also have to polish the knowledge and skills that robots do not yet possess. These include creativity, social skills, empathy, flexibility, problem-solving, asking questions, cooperation, thinking in terms of systems, and the ability to continue learning and to change. The more of these skills a job requires, the more difficult it will be to automate.

Third, robotisation will raise new questions about the distribution of wealth and income. Went: 'Some groups of people will see their incomes increase, while others will lose income. I think that we should be generous to the groups who lose out even with re-training, and we should offer them some form of financial security. We'll have to keep our finger on the pulse of the changes in wealth and income.'

Finally, robots should supplement humans, not replace them. Went: 'Don't treat people like robots, and make robots that help people do their work better. Let people be in charge of their own work and that of the robot. That will help them feel happiest, and they will be the most productive. Technology is supposed to serve humans, not the other way around.'

Fusing mind and body with soft robotics

Using technology to improve ourselves is not such a strange idea. Contact lenses are a simple way to help far-sighted or near-sighted people see clearly, but lenses can also be used in microscopes and telescopes. They help us see things that are too small to be seen by the naked eye, as well as galaxies that are millions of light years away.

Prostheses are sometimes even better than the human limbs they replace. Runners with carbon fibre blades, such as the South African sprinter Oscar Pistorius, can run as fast or faster than humans running on their own legs. In a TED presentation, the American athlete and actress Aimee Mullins explained that she doesn't feel hindered or handicapped by the fact that she doesn't have lower legs. In fact, she feels like a kind of superhero, because she can switch between different pairs of artificial legs, depending on her needs and her mood at the time. The fashion designer Alexander McQueen made her a pair of hand-carved legs from ash wood, and she was fitted with a pair of transparent lower legs for a role in a film. Mullins also has prosthetic legs of different lengths in her wardrobe, so that when she goes to a party, she can choose to show up 10 centimetres taller than usual. Her missing lower legs are a kind of blank canvas that she can fill any way she chooses.

If it's already possible to create these kinds of options with fairly simple artificial limbs, then robotics can offer even more possibilities. The idea of becoming one with technology may seem frightening at first — wouldn't that make me a cyborg? — but you may have already taken several steps in that direction without even realising. Anyone who has gone without a smartphone for a few days has noticed how accustomed we've become to living with a computer in your pocket. Only then do you realise how often you contact people when you're out and about — What is your address again? Shall I bring a takeaway? I'm running a bit late! — and how handy it is to be able to see if you're still on the right track when you're going somewhere you haven't been before.

The feeling of your body or mind becoming one with a lifeless object is more common than you might think. When you first try to ride a skateboard or use skates, you probably feel unstable and uncoordinated. But the more you practice, the more you feel one with your skateboard or skates. When you start cycling at a young age, you sense precisely how fast you can go around a corner, how sharp you can turn the corner, and how to stay perfectly in balance at low speeds. People who sit in a wheelchair or walk using an exoskeleton also start to feel as if the apparatus is part of their own body after a while.

In a certain sense, we are all therefore cyborgs, fused to our smartphones, computers, and bicycles — with the important difference that we can easily set aside our devices. And maybe it wouldn't be so bad to become a cyborg. In books and films, cyborgs are generally beings with super-human strength, who can run faster, fight better, and jump higher than humans. If technology can help us to function better, why should we deny ourselves that power? It's not such a strange idea to gradually replace parts of our body piece-by-piece with robotic components, if they function better than our own 'biological' body parts. Faster legs, stronger hands, sharper eyes, a more sensitive nose, a more beautiful singing voice, whiter teeth...

Smart materials — materials that can change characteristics, such as colour or rigidity, on command — can help make these improvements possible. By using smart materials, we may be able to build exosuits: soft robot suits that can help us to remain more mobile for longer periods. These types of robot suits may also be able to make us stronger and faster than we would be without them.

Scientists are also working on soft robotic materials that can be implanted into the body to repair or replace damaged organs. Experts predict that we will begin to see these 'bio-integrated' soft robotics in hospitals within the next 10 to 15 years. Smart materials will also make it possible to build robots that dissolve completely once they've done their work inside the body, such as delivering artificial tissue to a damaged organ. In that sense, robotics is approaching the field of chemistry: degradable molecules that the body doesn't reject make it possible to develop these kinds of 'robotic drugs'.

Another area of robotics that involves chemistry is the field of nanorobotics. The smaller a robot becomes, the closer you get to tinkering at the molecular level. Chemist Ben Feringa, Professor at the University of Groningen, won the Nobel Prize in 2016 for his work on 'molecular motors'. A molecular motor — also referred to as a 'nanomachine' — is only a few nanometres across and moves when exposed to ultraviolet light. In the future, nanomachines could be used to deliver medications to specific locations in the human body.

We don't consider nanomachines and smart materials to be true robots as such, but they may represent the future of robotics: not just robots that look like the classic ideal of a metal machine, but also clothing and medicines that have an almost invisible robotic component. In so doing, technology and humans can combine to create a better, healthier human body.

And then the smart robot became creative

It's 2035 in the science fiction film *I, Robot* (2004). Police detective Del Spooner is interrogating the humanoid robot Sonny about the apparent suicide of a famous robot designer. During the investigation, Spooner says to Sonny: 'Human beings have dreams. Even dogs have dreams. But not you. You are just a machine. An imitation of life. Can a robot write a symphony? Can a robot turn a canvas into a beautiful masterpiece?'

To which Sonny replies: 'Can you?'

Indeed, as if every human were so creative.

Since we don't really know how the creative process works exactly, we have a tendency to mystify the phenomenon. The ancient Greeks thought that creativity had divine origins. But during the go match between the computer AlphaGo and the human world champion Lee Sedol in 2016, the computer made a brilliant move that enriched the entire game. At the moment the computer made the move, though, all of the go analysts thought it was dumb. No go champion would ever think to make such a move. Only after a deeper analysis did they realise that the computer was completely right.

Was that move really creative?

That depends on what how we define 'creative'. Most definitions of creativity involve at least two elements: the product of creativity must be new, and it must be in some way useful, applicable, or valuable. The invention of the paperclip resulted in something new, and it was useful, so we consider the paperclip to be a creative invention. Artists create work that is new and that is considered to be of cultural value, so we see their work as creative as well.

Sometimes an extra requirement is added to the definition of creativity, in that the product must not only be new, but also 'out of the box'. With this extra requirement, a child's drawing of a smiling sun is not creative, because many children draw the same thing. But Einstein's theory of relativity is creative, because it required him to think 'outside the box'. He had to consider gravity to be a characteristic of space and time combined.

The brilliant move by AlphaGo was definitely new and useful within the context of the game of go. It was also 'out of the box', in that humans would never have made the move, because they thought that the history of the game had taught that it was a bad choice. So yes, according to this definition, AlphaGo did indeed make a creative move.

Since then, robots have composed and performed music, and there are computer programs that can write prose and poetry. At the website botpoet.com, you can even try to guess whether a poem was written by a human or by a computer, which can be very difficult to do. Not that the computer understands what it has written, or that it wrote the poem with a specific intention in mind, but the end product can seem very similar to a poem written by a human.

In fact, many people guess incorrectly that the poetry by the American author Gertrude Stein was written by a computer, because she apparently gives a machine-like twist to her poems: 'Rose is a rose is a rose is a rose.'

The British painter Harold Cohen, who passed away in 2016, was the creator of the painting robot Aaron. Humans can't tell whether Aaron's paintings were created by a human or a robot. Cohen himself thought that his robot was creative with a lower-case 'c', rather than a capital 'C'. By that he meant that Aaron was programmed using a collection of rules, to which it added random variations. The computer pioneer Ada Lovelace described such an approach in the 19th century: If a machine can only do what it is programmed to do, how can it ever be described as creative?

Aaron can create interesting new paintings, but it will never come up with entirely new rules and fundamentally change its own computer program. A painter like Picasso, who developed entirely new styles of painting, can do what Aaron cannot, which is why Cohen called Picasso's creativity 'creativity-with-a-capital C'.

Yet there is no fundamental reason why a robot could not become 'creative-with-a-capital-C'. Creativity often occurs when someone unconsciously associates two completely different ideas, and uses them to make something entirely new. Like humans, robots can conduct experiments in the real world, acquire new experiences, and eventually combine existing ideas together into something that is truly 'out of the box', such as a revolutionary new style of painting or theory of physics. It may not happen tomorrow, but it's certainly not impossible.

However, a major difference between the game of go and the world of physics, for example, lies in the range of possible solutions. In the game of go, the range of possible solutions is exactly delineated. The number of possible moves is unbelievably large, but it's still finite, and all of the possible moves are known. In the field of physics, however, the range of possible solutions is virtually unlimited. How did Einstein come up with the idea of combining time and space, and to consider gravity as a characteristic of space-time? That's a mystery.

In physics, like in everyday life, the range of possible solutions to a problem are limited only by physical reality, plus the entire conceptual and imaginary world inside a person's head. That makes it only unimaginably large, but also difficult to describe in rules. Humans have learned through experience how to search through those possibilities in order to solve a problem. However, if a robot were to acquire sufficient experience, collect sufficient data, and have access to sufficiently effective learning models, then in principle it should be able to learn from its experiences and find creative solutions for new problems the same way that humans do.

At the University of Manchester, Professor of Computer Science Ross D. King has built the robotic scientist Eve. Eve is equipped with several robotic arms, cameras, sensors, and pipettes. The robot conducts fully automatic scientific experiments on yeast cells, which she removes from the freezer on its own. Eve also generates hypotheses about the functions of genes inside the yeast cell, then she designs an experiment to test that hypothesis, conducts the experiment, analyses the results, and can even suggest follow-up research. 'For example, Eve has already found a substance that may be able to serve as an anti-malaria drug', King explains in a Skype interview.

King is working to take the research process used by pharmaceutical companies to discover new drugs and roboticise it as much as possible. 'It's actually easier for robots to perform science than it is for them to create art, because science involves a lot of formal reasoning, and robots can do that better than humans.

Moreover, when humans conduct experiments, sometimes they see what they want to see. If it's properly programmed, a robot will only see what's really there.'

At the moment, Eve still needs a lot of human supervision. And King explains that the added value robots provide in creative processes such as science, the arts, and architecture, will mainly lie in supplementing human creativity, although their contributions will become richer over time. 'When it comes to creativity, I think that there is a spectrum ranging from what a robot like Eve can already do, to the creativity that you and I possess, to the creative genius of Einstein or Picasso. Step-by-step, robots will become more creative. I am convinced that in principle, robots can do anything humans can do, and eventually they will be able to do much more.'

Will robots really take over the world?

Will robots take over the world? Should we be afraid of super-intelligence? Computers can already beat us at chess and in all kinds of computer games. How long will it take before they take over from us in the real world?

The first step in that direction is perhaps a real human robot, a robot that is 'someone' instead of 'something,' who has free will and frees himself from his maker. But can a robot ever really develop consciousness? And how do you actually find out? After all, you can already program a computer in such a way that it responds to the question 'Do you have consciousness?' with a convincing 'Yes.'

Philosophy, and thinking about consciousness, has existed for thousands of years. Thinking about the question of whether computers and robots have a consciousness, has gone on for a comparatively short time. Yet Alan Turing, mathematician and founder of computer science, wondered in 1950: can machines think? He actually found that question 'too meaningless to discuss,' because how can you actually find out if machines can think, and what do we mean by 'thinking' at all?

Because it is difficult to know from outside whether something can think, Turing proposed an imitation game to replace that question. In this game, which we now call the Turingtest, a human being and a computer both pretend to be human and a subject has to find out which is a computer by talking to them. Turing predicted that in the year 2000 we would already have computers that can play the imitation game so well that an average interrogator would not have more than 70 per cent chance of making the right choice after a five-minute interrogation.

Depending on how you interpret Turing's short explanation, his test can be carried out in different ways. In 2014, a robot passed one such Turing test: the chatbot Eugene was able to convince just enough jurors that it was a human being. Yet if Turing had been able to watch, he would undoubtedly have adjusted his requirements for the test. The chat program had been given the personality of a 13-year-old Ukrainian boy, so that the jury would not set its requirements too high, at least not as high as in adults with English as their mother tongue.

Turing died in 1954 and was therefore unable to see most developments in this area. Nevertheless, in 1950 he predicted that there would be a lively discussion about whether robots could become human. He set out the nine most important arguments against thinking machines. The theological objection, for example: God has given people a soul, and that soul ensures that we — and therefore not machines or animals — can think. Or the 'head-in-the-sand' objection: we find man superior to all other things, and therefore we cannot imagine that a machine could also be human.

He called one the possible counter-arguments the conscious-ness objection. You can never know for sure if another person has consciousness, Turing said, but with people we quickly assume that they have consciousness when we deal with them at a sufficiently advanced level. If you have just such advanced interaction with a machine — if that machine passes the Turingtest — there is no reason to deny that the device has a consciousness.

Turing also mentions the 'argument of disabilities.' This is actually a collection of arguments that all look like 'I admit you can make machines that can do everything you've just described, but you will never be able to make a machine that X can do.' That X can be anything, but Turing mentions a selection: 'Be kind; be resourceful; be beautiful; be friendly; have initiative; have a sense of humour; tell right from wrong; make mistakes; fall in love; enjoy strawberries and cream; make someone fall in love with one; learn from experience; use words properly; be the subject of one's own thoughts; have as much diversity of behaviour as a man; do something really new.'

The reason why people think so, says Turing, is because they have too little imagination to imagine something so new. The first robot to enjoy strawberries and cream has yet to be made, but it is not impossible that this will happen in the foreseeable future. Yet this is a weak argument: in the decades since Turing's article we have already been able to tick off some of the list of disabilities, but you can endlessly supplement the list with new things.

Yet many people fear that robots will become so smart that they will take over the world. Thus the physicist Stephen Hawking warned that the development of artificial intelligence might mean the end of humanity. Elon Musk, founder of Tesla electric cars, has said that man could become no more than a pet of super-intelligent robots.

But neither Hawking nor Musk are experts in the field of robot-ics and artificial intelligence. A group of experts concluded in the authoritative research 'One hundred year study on artificial intelligence (AI100)': 'Contrary to the more fantastic predictions for AI in the popular press, the Study Panel found no cause for concern that AI is an imminent threat to humankind. No machines with self-sustaining long-term goals and intent have been developed, nor are they likely to be developed in the near future.'

Artificially intelligent computers and robots are already very good at some tasks. But 'artificial general intelligence' — a computer or robot that can do everything that people can — is still far away. The step from a robot that can walk, talk, play chess, or drive a car to real intelligence, let alone super intelligence, is immense.

According to most roboticists, the construction of super-intelligent robots that pose a threat to humanity is theoretically possible, but practically unlikely. Roboticist Alan Winfield puts it this way: 'If we succeed in building human equivalent AI and if that AI acquires a full understanding of how it works, and if it then succeeds in improving itself to produce super-intelligent AI, and if that super-AI, accidentally or maliciously, starts to consume resources, and if we fail to pull the plug, then, yes, we may well have a problem. The risk, while not impossible, is improbable.'

The American science writer Michael Shermer, founder of The Skeptics Society, adds: why should artificial intelligence actually want to take over the earth? A super-intelligent robot does not have to behave like an alpha male that wants to crush competitors — it may well be a peace-loving being that wants to solve all our problems for us. Shermer says: 'Given the zero per cent historical success rate of apocalyptic predictions, coupled with the incrementally gradual development of AI over the decades, we have plenty of time to build in fail-safe systems to prevent any such apocalypse.'

Robot timeline
BEFORE 1920

FICTION

9th/8th century B.C. Automatons (*The Iliad*, Homer) — epic poem
In the epic poem *The Iliad*, Homer describes the first 'automatons': autonomous machines created by the divine blacksmith Hephaestus and the Athenian craftsman Daedalus.

Pygmalion's statue — Greek mythology
According to the ancient Greek myth, Pygmalion created an ivory statue of a woman that was so beautiful, he fell in love with his own creation. Aphrodite transformed the statue into a real woman, and the two lived happily ever after. The myth was recorded by several writers, including the Roman poet Ovid in his *Metamorphoses*.

Golem - Jewish legend
In the Jewish legend, the Golem is a creature made out of clay and brought to life by a rabbi.

1818 Frankenstein's Monster (*Frankenstein*, Mary Shelley) — novel

The scientist Victor Frankenstein builds a creature out of human body parts and brings it to life. The creature escapes and develops murderous tendencies. A few decades later, Isaac Asimov coined the term 'Frankenstein complex' to describe our innate fear of robots, along with the technophobic world view that he despised.

1883 Pinocchio (*The Adventures of Pinocchio*, Carlo Collodi) — novel
The woodcarver Geppetto is given a block of living wood that he uses to carve a marionette in the form of a boy: Pinocchio.

FACT

400 B.C. The Greek scientist Archytas builds a wooden automaton in the form of a dove. Automatons like The Dove are the predecessors of modern robots.

c800 A.D. Three Persian brothers known as the Banū Mūsā are commissioned by the Caliph of Baghdad to write The Book of Ingenious Devices, in which they also describe a number of automatic machines.

Early 13th century Ismail Al-Jazari of Persia builds a simple robot orchestra with four musicians.

1497 Leonard da Vinci designs a robot knight. Some years later, he demonstrates a mechanical lion to the King of France.

17th/18th century Japanese craftsmen build automatic mechanical dolls, called karakuri. The dolls were able to shoot a bow and arrow or serve guests a cup of tea.

c1750 Friedrich von Knauss builds automatons that can play a musical instrument and write short sentences.

1769 Wolfgang von Kempelen builds The Turk, a fake chess machine that hides a human chess player.

1805 Joseph-Marie Jacquard develops a method for controlling looms via punch cards. English textile workers protest in fear of losing their jobs.

1898 Nikola Tesla demonstrates a model for a remote-controlled submersible.

1920 - 1929

FICTION

1920 First use of the word 'robot' to describe a machine *(Rossum's Universal Robots*, **Karl Čapek) — stage play**
The Czech writer Karel Čapek uses the word 'robot', derived from the Czech word for forced labour: 'robota'. The robots, which are organic humanoids, are perfect labourers that work in factories, offices, and even the army. Humans, left with nothing to do, become lazy and even stop reproducing. As the robots begin to acquire more human traits, they rise up against their human oppressors.

1921 *The Mechanical Man (L'Uomo Meccanico*, **André Deed) — film**
The first fight between two robots on the silver screen. A gang of thieves steals a robot and uses it to commit all sorts of crimes. Part of the film is lost, but the rest has been preserved and is now part of the public domain, and therefore legally available on YouTube.

1927 Maschinenmensch/Robot Maria (*Metropolis*, **Fritz Lang) — film**
In a dystopian future, the vast majority of humanity lives and works underground in order to support the lives of the happy few above ground. A mad professor builds a robot inspired by his deceased love, the girl Maria, and brings it to life. The 'Mashinenmensch' provokes the underground workers to protest, which causes all of the pump equipment to come to a standstill, threatening to drown the underground city.

FACT

1929 The first mechanical robots that resemble humans are exhibited at fairs in England and the US.

1930 - 1949

FICTION

1939 The Tin Man (*The Wizard of Oz*, Victor Fleming) — film
Dorothy and her dog Toto are carried away by a tornado to the magical land of Oz. There, they meet The Tin Man, a robot-like metal man whose greatest wish is to have a heart.

1942 Speedy (*Runaround*, Isaac Asimov) — short story
The first mention of the Three Laws of Robotics. The robot Speedy is ordered to remove selenium from a pool, but recoils from the harmful gases that have accumulated around the pool. As Speedy is an extremely expensive robot, its makers have strengthened the Third Law of robotics in order to prevent Speedy from endangering itself.

As Speedy approaches the selenium pool, he senses a conflict between the Third Law, which commands it not to get too close to the pool, which would put it in danger, and the Second Law, which requires it to follow the human command to remove the selenium.

FACT

1939 At the New York World's Fair, a two-metre tall mechanical humanoid robot named Elektro is put on display. Elektro can walk, talk (700 words), and smoke. In 1940, Elektro is joined by the robot dog Sparko.

1942 The first patent for a programmable paint spray mechanism is granted.

1948 William Grey Walter and his wife Vivian build the first mobile robots, the electromechanical Elmer and Elsie, nicknamed 'the turtles'.

1949 The EDSAC computer built by Maurice Wilkes in Cambridge is the world's first computer able to store a computer program in its own memory. The development of the first electronic computers during and after World War II make it possible to build programmable robots for the first time.

FICTION

1951 Gort (*The Day the Earth Stood Still*, Robert Wise) — film
The alien Klaatu and his robot Gort bring a peaceful message to Earth.

1952 Astro (*Astro Boy*, Osamu Tezuka) — manga
The robot boy Astro reveals itself to be a true superhero with super-human powers. After the first series of mangas (Japanese comic strips) in the 1950s, Astro also appeared in several animated series, a live-action film, an animated film, and several computer games.

1954 Tobor (*Tobor the Great*, Lee Sholem) — film
Concerned that space travel is too hazardous for humans, researchers build the humanoid metal robot Tobor. But before Tobor can be sent to space, the inventor and his grandson are kidnapped, and Tobor hurries to the rescue.

1956 Robby the Robot (*Forbidden Planet*, Fred M. Wilcox) — film
Dr. Edward Morbius and his daughter Altaira live on a distant planet together with their robot Robby. Robby is not only a fully-fledged character, but also a source of comic relief.

The robot suit becomes a real movie star: it is later used for other films and TV series, including *The Invisible Boy*, *The Addams Family* and *Lost in Space*.

FACT

1951 The Frenchman Raymond Goertz develops the first remote-controlled robot arm.

1953 The first remote-controlled submersible robot is used for underwater photography.

1954 George Devol develops the first programmable robot arm.

1956 Dartmouth Summer Research Project on Artificial Intelligence, the conference that is generally considered to be the starting point for artificial intelligence.

1960 - 1969

FICTION

1962 Rosie (*The Jetsons*, Hanna-Barbera) — TV series
Rosie is the Jetson family's household robot in the futuristic world of the year 2062. She cooks, cleans house, and cares for the children.

1963 Daleks (*Doctor Who*, Christopher Barry/Richard Martin) — TV series
The Daleks are robotic villains perhaps best known for their chilling catchphrase 'Exterminate!' They are extremely aggressive: 'However you respond is seen as an act of provocation', in the words of the protagonist, The Doctor.

1965 Trurl and Klapaucius (*The Cyberiad*, Stanislaw Lem) — novel
In a world somewhere between the distant future and the dark ages, robot builders take the place of wizards and magic. One day, the robots Trurl and Klapaucius build a machine that can make anything that starts with the letter N.

1968 HAL 9000 (*2001: A Space Odyssey*, Arthur C. Clarke/Stanley Kubrick) — film
HAL 9000 doesn't have a classic robot body; instead, its body is the spaceship Discovery, on its way to the planet Jupiter. It does not show emotions, but it definitely has its own will. In fact, its will is strong enough that it can refuse one of the astronauts, Dave, access to the ship after a spacewalk with the famous words: "I'm sorry Dave, I'm afraid I can't do that." HAL 9000's red camera-eye has become an icon in the world of science fiction.

FACT

1961 The father of robotics, Joseph Engelberger, founds the company Unimation Inc. The company sells the robot arm Unimate, developed by George Devol.

1961 General Motors puts the first industrial robot to work: robot arm Unimate is used to spray-paint cars in the automotive industry. In the decades that follow, the automotive industry is the driving force behind the development of robots.

1965 Artificial intelligence pioneer Herbert Simon predicts that within 20 years, machines will be able to do all of the work that humans can do.

1966 The mobile robot Surveyor lands on the moon.

1967 The Swedish firm Svenska Metallverken (ABSM) is the first European company to install an American Unimate robot.

1968 Artificial intelligence pioneer Marvin Minsky at the Massachusetts Institute of Technology (MIT) develops an arm with 12 joints: the Tentacle Arm.

1969 Shakey is the first mobile robot that can think about its own actions and drive around a room.

1970 - 1979

FICTION

1973 The Gunslinger (*Westworld*, Michael Crichton) — film
In the futuristic year of 1983, visitors to the amusement park Westworld can live out their wildest fantasies – mainly sex and violence – on android robots. In 2016, an eponymous American television series based on the film was released.

1973 Steve Austin (*The Six Million Dollar Man*) — TV series
When astronaut Steve Austin is injured in an accident, he is repaired using robotic components costing — you guessed — $6m.

1976 Andrew (*The Bicentennial Man*, Isaac Asimov) — novel
The Martin family purchases a humanoid robot for household chores. The robot, Andrew, develops creativity and a sense of humour, and decides to buy his own freedom and develop as a human.

1977 C-3PO and R2-D2 (*Star Wars*, George Lucas) — film
R2-D2 is an astromech droid and C-3PO is fluent in six million forms of communication. The robots are the only characters portrayed in each of the first six Star Wars films to be played by the same actor.

1979 Marvin the Paranoid Android (*Hitch-hiker's Guide to the Galaxy*, Douglas Adams) — novel
The humanoid robot Marvin is actually a failed prototype for artificial intelligence, built against its own will. Marvin has a brain the size of a planet, but never has the opportunity to utilise its enormous intellectual abilities. As a result, it suffers from chronic depression.

FACT

1973 The Tomorrow Tool (T3) is the world's first industrial robot to be controlled by a miniature computer.

1973 The Japanese Ichiro Kato builds the first humanoid robot: Wabot-1.

1973 Robot Freddy II, built in Edinburgh (UK) under the supervision of Donald Michie, can assemble a simple toy car or boat from a number of individual parts.

1973 The German firm KUKA stops importing Unimate robots from the US and begins developing its own robots.

1973 The Japanese firm Hitachi develops the first robot that can twist bolts.

1976 Robot arms on the Viking 1 and 2 rovers take soil samples on Mars.

1977 Researchers at the Russian Academy of Sciences build the six-legged walking robot Variante Masha.

1979 Hans Moravec builds the Stanford Cart, an autonomous vehicle that can drive around a room.

1979 An industrial robot operated by the Ford Motor Company kills an American engineer. This is the first known fatal accident caused by a robot.

1980 - 1989

FICTION

1982 Replicants (*Blade Runner*, Ridley Scott) — film
Replicants are genetically modified android beings that are identical to humans, except for their lack of emotions. The Voight-Kampff test measures whether the subject has emotions, and can therefore differentiate humans from replicants. But what if the test shows that someone doesn't have emotions, even though they are absolutely convinced they are human?

1982 KITT (*Knight Rider*, Glen A. Larson) — TV series
Crime hunter Michael Knight is assisted by his intelligent car KITT, which can think, talk, and perceive its surroundings. A battery of red lights blink back and forth on its front bumper. Interestingly, KITT needs a human driver. The idea that an intelligent car could drive itself was too futuristic in the early 1980s.

1983 Inspector Gadget (*Inspector Gadget*) — animated series
Inspector Gadget is a clumsy police inspector who has a variety of bionic parts. 'Go, go, gadget arms!' he commands — and his arms shoot out.

1984 The Terminator (*The Terminator*, James Cameron) — film
Skynet is an intelligent computer program that wants to eliminate humanity with the help of a cyborg that looks like a human: The Terminator, played by Arnold Schwarzenegger.

1984 Autobots and Decepticons (*Transformers*) — toys
The Transformers are humanoid robots that can transform into cars or other vehicles. They were initially brought out as toys, but resulted in spin-off comic books, films, novels, and games.

1987 RoboCop (*RoboCop*, Paul Verhoeven) — film
When police agent Alex Murphy is killed, he is rebuilt as a cyborg. In his new identity, he has three prime directives: to serve the public trust, to protect the innocent, and to uphold the law.

FACT

1980 The American firm Automatix produces the first industrial robots with automatic image recognition.

1984 Valentino Braitenberg describes analogue 'vehicles' with a simple structure, but which display complex behaviour. These Braitenberg vehicles are a major source of inspiration for roboticists.

1986 The company Redzone Robotics begins production of robots intended specifically to work in hazardous environments.

1989 Rodney Brooks and Anita Flynn publish the article 'Fast, Cheap and Out of Control', which leads to the development of small, relatively unintelligent robots, instead of complicated humanoid robots. One example is the six-legged insectoid robot Genghis, also from the year 1989.

1989 Danbury Hospital in Connecticut (US) is the first to use a service robot: HelpMate can bring trays of food to patients.

1990 - 1999

FICTION

1995 Major Motoko Kusanagi (*Ghost in the Shell*, Masamune Shirow) — manga
Motoko Kusanagi is a cyborg, living in a full-body prosthesis after a fatal accident as a child.

1995 Golems (*Discworld*, Terry Pratchett) — book
Like the Jewish legend, these Golems are also a kind of robot made out of clay. They explicitly obey Asimov's Three Laws of Robotics, and consider themselves to be property until they can buy their freedom.

1996 The Borg (*Star Trek: Voyager*) — TV series
The Borg is a community of cyborgs led by a Borg Queen. They travel through space looking for new beings to assimilate. One of the Borg, Seven of Nine, is liberated from her Borg implants and joins the crew of the Voyager, becoming one of the most important characters in the TV series.

1999 Bender (*Futurama*, Matt Groening) — TV series
Bender looks like a fairly standard robot, but he doesn't behave entirely as one would expect from a robot. He uses alcohol as fuel, smokes, steals, gambles, curses, chases women, and generally does everything God forbids.

1999 Sentinels (*The Matrix*, The Wachowskis) — film
What we see as reality is actually a simulation. A small band of rebels manages to break out of the simulation and searches the real world for a way to shut the simulation down. In the process, they are threatened by Sentinels, autonomous flying killer robots in the form of gigantic squid.

FACT

1990 The robot arm Robodoc is used to perform a hip implantation on a dog. Two years later, Robodoc is used on a human for the first time, also for a hip implant.

1993 Rodney Brooks, Lynn Stein and Cynthia Breazeal begin building COG, in the hope of being able to build a robot that behaves like a two-year-old child within five years. That proved to be far too ambitious.

1994 Two robot cars, the VaMP and the VITA-2, carry passengers for a thousand kilometres through ordinary traffic on European motorways.

1995 First use of the Predator military drone. The Predator is controlled remotely from a base inside the United States.

1996 Honda demonstrates the humanoid robot Prototype 2 (P2), which can walk, climb stairs, and carry objects.

1997 The first RoboCup tournament for football-playing robots is held in Japan.

1997 The robot rover Sojourner lands on Mars.

1997 NASA begins development of Robonaut, a humanoid robot without legs, which will help future astronauts perform tasks outside spaceships and space stations.

1998 Tiger Electronics introduces Furby, the first interactive robotic pet.

1999 Sony introduces the robotic toy dog Aibo to the market.

2000 - 2009

FICTION

2001 David (*A.I. Artificial Intelligence*, Steven Spielberg) — film
When their son is seriously ill, Monica and her husband Henry decide to buy the robot boy David. But when their son comes home from the hospital, the two boys have a fight, and Monica and Henry decide that David should be destroyed. On the way to the factory, Monica has second thoughts, and leaves David alone in the woods.

2004 Sonny (*I, Robot*, Alex Proyas) — film
During an investigation, police agent Del Spooner is attacked by the robot Sonny – which is officially forbidden, as all robots must obey Asimov's Three Laws of Robotics. Sonny is actually under the command of the artificial superintelligence VIKI, which threatens to take away humanity's free will.

2005 Rodney Copperbottom and other characters (*Robots*, Chris Wedge) — film
In the cheerful animated film Robots, the entire cast are robots.

2007 GLaDOS (*Portal*) — game
GLaDOS – Genetic Lifeform and Disk Operating System – is the narrator of the computer game Portal. The feminine robotic voice leads the player through the first levels of the game. But then she turns against the player… GLaDOS won several awards as the best new game character and the best villain in a video game.

2008 WALL-E and EVE (*WALL-E*, Andrew Stanton) — film
WALL-E is a rusty, cube-shaped robot that cleans up rubbish on an abandoned, heavily polluted Earth. One day, the elegant, spotless robot EVE lands on Earth to investigate whether the planet has become habitable again. A robot romance develops between the two.

FACT

2000 Honda introduces the humanoid robot Asimo, the successor to the P2 from 1996.

2000 Cynthia Breazeal (MIT) develops Kismet, a robot face that can express emotions. It is a pioneering work in the field of social-emotional interaction between humans and robots.

2001 First robot operation: surgeons in London conduct prostate surgery by controlling three robot arms with joysticks.

2001 Search and rescue robots, such as the remote-controlled PackBot by iRobot, are used at Ground Zero after the terrorist attack on the World Trade Center buildings in New York.

2002 iRobot begins sales of its robot vacuum cleaner Roomba, the most successful robot for consumer use to date.

2002 First use of the military robot PackBot in Afghanistan.

2004 DARPA organises the first Grand Challenge, a contest for robot cars. None reaches the finish line.

2004 Start of the EU project RobotCub: the one-meter tall humanoid robot iCub is built as a test platform for research into human cognition and artificial intelligence. iCub resembles a two-and-a-half-year-old toddler, and can see, hear, and make advanced movements.

2005 Under the leadership of Sebastian Thrun, Stanford Racing Team's robot car Stanley wins the second DARPA Grand Challenge. Five of the 23 participating cars reach the finish line after 212 kilometres.

2005 Hod Lipson from Cornell University (US) builds robots out of blocks that can display a simple form of self-replication.

2006 The French company Aldebaran Robotics builds the interactive, programmable, and affordable humanoid robot Nao, an ideal robot for research and education.

2010 - 2017

FICTION

2012 Robot (*Robot & Frank*, Jake Schreier) — film
The aging former criminal Frank is given a care robot, which he calls simply 'Robot'. Frank is initially suspicious of Robot, but he gradually begins to feel affection for it, and uses it to help him commit one last jewel heist.

2012 Hubots (*Äkta Människor/Real Humans*) — TV series
Humans live in harmony with a wide range of Hubots; android robots that are virtually indistinguishable from humans. As the Hubots begin to become accepted as equals in society, a small group of people forms a political party against robots and robot rights.

2014 Baymax (*Big Hero 6*, Don Hall/Chris Williams) — film
The boy Hiro and his big brother Tadashi build robots, including a robot swarm and the large inflatable care robot Baymax.

2014 TARS and CASE (*Interstellar*, Christopher Nolan) — film
TARS and CASE are intelligent robot assistants travelling along on a secret space expedition to find habitable planets in a distant solar system.

2015 BB-8 (*Star Wars: The Force Awakens*, J.J. Abrams) — film
Almost 40 years after the first Star Wars film, the iconic robots R2-D2 and C-3PO get a new mate: the white-and-orange BB-8, a ball-shaped robot with a domed head on top.

2015 Ava (*Ex Machina*, Alex Garland) — film
The wealthy Nathan invites Caleb to spend a week at his remote estate for a week. Over the course of the week, Caleb starts a relationship with the beautiful robot Ava, who was built by Nathan. A conflict arises between Ava, Caleb and Nathan.

FACT

2012 Google demonstrates the first self-driving car: Google Car.

2015 Robot Pepper is introduced to the market: an interactive humanoid robot that can communicate both by voice and by hand gestures. Some companies use Pepper to welcome guests.

2016 First fatal accident caused by a self-driving car, a Tesla S.

2016 The Dallas Police Department in the United States uses a tracked 'bomb robot' to kill a shooter who had just murdered five police agents.

2016 Guszti Eiben from VU Amsterdam has two robots 'mate' with one another, using a 3D printer to give birth to the world's first robot baby; a proof of principle that robots are capable of reproduction.

2016 Researchers at Harvard University in America develop a biohybrid robot ray that is 1.5 centimetres long. The robot ray can swim using a silicon body with live rat heart cells as muscles.

2017 Automotive manufacturer Tesla introduces the Tesla 3, an affordable electric car that is equipped with everything it needs to be able to drive autonomously in the future: seven cameras, a radar, nine distance sensors, and a computer.

Resources and reading material

Sources and reading material
Links to the websites, articles, videos, and other reference materials can be found at: hallorobot.nl/references

Chapter 1

Interviews
David Hanson (9 June 2017; AI for Good conference in Geneva), Pep Rosenfeld (13 July 2017)

Literature and other media
'Meet Mr. Robot — Not Forgetting His Master', *The Age*, 20 September 1935
J.S. Brown, 'Toy Automaton', U.S. Patent 40891, 8 December 1863
Helen Greiner, 'Time for robots to get real', *New Scientist*, 18 January 2012
Erico Guizzo, 'Hiroshi Ishiguro: The Man Who Made a Copy of Himself', *IEEE Spectrum*, 2010
Bennie Mols, '(Kunst)matige intelligentie', VPRO Gids, 6 June 2015
Michael E. Moran, 'The da Vinci robot', *Journal of Endourology*, 2006
David Whitehouse, 'Japanese develop "female" android', *BBC News*, 27 July 2005
The official website of the RoboCup Humanoid League
'Bina 48 meets Bina Rothblatt — Part One', The LifeNaut Project, via YouTube

Chapter 2

Interviews
Laurens van der Maaten (25 juli 2016), Tom Rijndorp (10 July 2017)

Literature and other media

Zachary Davies Boren, 'Robot With Artificial Tongue Can Appreciate Fine Wine Better Than Humans', *International Business Times*, 23 September 2014

Louise Dewast, 'Government Robots Will Decide If Your Thai Food Tastes Right', *ABC News*, 30 September 2014

Rik Nijland, '*Niet langer alleen op de ogen vertrouwen*', *Wageningen World*, 31 May 2011

Seymour A. Papert, 'The Summer Vision Project', Memo AIM-100, *MIT AI Lab*, 1966

Malika Sevil, '*Elektronische neus van AMC ruikt longziekte*', *Het Parool*, 5 July 2017

Benjamin C.K. Tee et al, 'A skin-inspired organic digital mechanoreceptor', *Science*, 2015

Jianxiong Xiao et al., 'Sun database: Large-scale scene recognition from abbey to zoo', *IEEE Conference on computer vision and pattern recognition*, 2010

'SCRATCHbot — A Rat like Robot', Razor Robotics, via YouTube

MIT Scene Recognition Demo, MIT Computer Science and Artificial Intelligence Laboratory

'Quipt: Taming Industrial Robots', Madlab.cc

'Researchers improve automated recognition of human body movements in videos', *Phys.org*, 8 June 2015

'ICEA: SCRATCHbot (2006-2009)', Bristol Robotics Laboratory

'Place and scene recognition from video', University of British Columbia

Chapter 3

Interviews

Olivier van Beekum, Vroukje en Hannah (3 October 2016)

Literature and other media
Joseph Guinto, 'Machine Man: Rodney Brooks', *Boston Magazine*, 28 October 2014
Bennie Mols, '*Gered door een robot*', *NRC Handelsblad*, 8 June 2013
Bennie Mols, '*Totaalvoetbal op wielen*', I/O *Magazine*, September 2012
Regan Morris, 'Why the 'cute robots' don't work for Rodney Brooks', *BBC News*, 17 November 2015
Joanne Pransky, 'The essential interview: Rodney Brooks, Founder of Rethink Robotics', *Robotics Business Review*, 12 February 2015

Chapter 4

Interviews
Esben Østergaard (10 juli 2015), tour of VDL Nedcar (12 July 2017)

Literature and other media
Bob Malone, 'George Devol: A Life Devoted to Invention, and Robots', *IEEE Spectrum*, 2011
George Munson, 'The First Industrial Robot: Why It Failed', *Robotics Business Review*, 2012
Jeremy Pearce, 'Joseph F. Engelberger, a Leader of the Robot Revolution, Dies at 90', *New York Times*, 2015

Chapter 5

Interviews
Jo Luijten (14 September 2016), Paul Vogt (14 July 2017)

Literature and other media
Marcel van Engelen, Interview met Luc Steels, Vrij Nederland, July 2015
Joshua Hartshorne, 'Where are the talking robots?', *Scientific American*

Mind, 2011

Will Knight, 'AI's Language Problem', *MIT Technology Review*, 2016

Tyrus L. Manuel, 'Creating a Robot Culture: An Interview with Luc Steels', *IEEE Intelligent Systems*, 2003

Jean-Pierre Martens, '*Spraaksynthese: van tekst naar spraak*', *Kennislink*, 2008

Erica Renckens, '*Het luisterend oor van de computer*', *Kennislink*, 2016

Chapter 6

Interviews

Cynthia Breazeal (8 June 2017; AI for Good conference in Geneva), Frauke Zeller (29 July 2017)

Literature and other media

Joseph Bates, 'The Role of Emotion in Believable Agents', *Communications of the ACM*, 1994

Cynthia Breazeal and Rodney Brooks, 'Robot Emotion: a Functional Perspective', *Who needs emotions?*, 2005

Cynthia Breazeal, 'Emotion and sociable humanoid robots', *International Journal of Human-Computer Studies*, 2003

Rafael Calvo, Sidney D'Mello, Jonathan Gratch and Arvid Kappas, *The Oxford Handbook of Affective Computing*, Oxford Library of Psychology, 2014

Antonio Damasio, *Descartes' Error: Emotion, Reason, and the Human Brain*, Putnam Publishing, 1994

Eva Hudlicka, 'To feel or not to feel: The role of affect in human-computer interaction', *International Journal of Human-Computer Studies*, 2003

Marvin Minsky, *The Emotion Machine: Commonsense Thinking, Artificial Intelligence, and the Future of the Human Mind*, Simon & Schuster, 2006

Andrew Ortony, Gerald L. Clore and Allan Collins, *The Cognitive Structure of Emotions*, Cambridge University Press, 1988

Rosalind W. Picard, *Affective Computing*, MIT Press, 1997

Daniel J. Rea, James E. Young and Pourang Irani, 'The Roomba mood ring: an ambient-display robot', *Proceedings of the seventh annual ACM/IEEE international conference on Human-Robot Interaction*, 2012

David Hanson, 'Robots that "show emotion"', TED, 2009.

'A Dog Tail for Robots' (2013), University of Manitoba

'JIBO, The World's First Social Robot for the Home', IndieGogo

Chapter 7

Interviews

Ivo Broeders (13 June 2017), Amber Case (19 May 2017), Robin Murphy (28 August 2015), Leila Takayama (18 February 2017)

Literature and other media

Isaac Asimov, *I, Robot*, 1950

Lisanne Bainbridge, 'Ironies of Automation', *Automatica*, 1983

Patrick Collinson, 'Pilotless planes: what you need to know', *The Guardian*, 7 August 2017

'Robots', special edition of *De Psycholoog*, November 2015

'"Gods" Make Comeback at Toyota as Humans Steal Jobs From Robots', *Bloomberg*, April 2014

Bennie Mols, 'Hoe meer automatisering, hoe crucialer de mens', *New Scientist* (Dutch edition), November 2013

Matt O'Sullivan, 'The untold story of QF72: What happens when "psycho" automation leaves pilots powerless?', *Sydney Morning Herald*, May 2017

Chapter 8

Interviews

Herman van der Kooij (14 July 2017), Claudia Bosch-Commijs (4 August 2017)

Literature and other media

Elizabeth Blair, 'The Challenge Of "Big Hero 6": How To Make A Huggable Robot', *NPR*, 7 November 2014

Leah Burrows, 'The first autonomous, entirely soft robot', The *Harvard Gazette*, 24 August 2016

'DARPA Helps Paralyzed Man Feel Again Using a Brain-Controlled Robotic Arm', *Darpa.mil*, 13 October 2016

Thomas Geijtenbeek, Michiel van de Panne and Frank van der Stappen, 'Flexible muscle-based locomotion for bipedal creatures', *ACM Transactions on Graphics*, 2013

Elliot W. Hawkes, Laura H. Blumenschein, Joseph D. Greer and Allison M. Okamura, 'A soft robot that navigates its environment through growth', *Science Robotics*, 2017

Carmel Majidi, 'Soft robotics: a perspective—current trends and prospects for the future', *Soft Robotics*, 2014

Michael Wehner et al, 'An integrated design and fabrication strategy for entirely soft, autonomous robots', *Nature*, 2016

Huichan Zhao et al, 'Optoelectronically innervated soft prosthetic hand via stretchable optical waveguides', *Science Robotics*, 2016

BEAM Reference Library

Hugh Herr, 'The new bionics that let us run, climb and dance', *TED*, 2014

'Meet Soft Robotics, Inc.', 2017, via YouTube

Chapter 9

Interviews

Guszti Eiben (10 July 2017), Wessel Straatman (11 July 2017), Martijn Wisse (20 July 2017)

Literature and other media

Robert Full, 'Robots inspired by cockroach ingenuity', *TED*, 2002

Robert Full, 'Learning from the gecko's tail', *TED*, 2009

Robert Full, 'The secrets of nature's grossest creatures, channelled into robots', *TED*, 2014

Peter Menzel and Faith D'Aluisio, *Robo Sapiens — Evolution of a new species*, MIT Press, 2001

Chapter 10

Interviews
Serge de Beer (14 June 2017), Guido de Croon (28 July 2017)

Literature and other media
Angus Chen, 'Heads up for the gathering robot swarm', *Science News*, 2014

Michael Rubenstein, Alejandro Cornejo and Radhika Nagpal, 'Programmable self-assembly in a thousand-robot swarm', *Science*, 2014

RobotTrainer, Serge de Beer

'Designing Collective Behavior in a Termite-Inspired Robotic Construction Team', Harvard University, via YouTube

'Swarmanoid, the movie', Mauro Birattari, via YouTube

'Symbiotic evolutionary robot organisms (SYMBRION)', University of Bristol

'High-Speed Robots Part 1: Meet BettyBot in "Human Exclusion Zone" Warehouses, *WIRED*, via YouTube

Radhika Nagpal, 'Taming the swarm — Collective Artificial Intelligence', *TEDxBermuda*, 2015

'A Compilation of Robots Falling Down at the DARPA Robotics Challenge', *IEEE Spectrum*, via YouTube

Chapter 11

Interviews
Catelijne Muller (26 July 2017)

Literature and other media
Craig A. Anderson and Brad J. Bushman, 'Effects of violent video games on aggressive behavior, aggressive cognition, aggressive affect, physiological arousal, and prosocial behavior: A meta-analytic review of the scientific literature', *Psychological Science,* 2001

'Asilomar AI Principles', Future of Life Institute

Isaac Asimov, 'The Three Laws', *Compute! Magazine,* Issue 18, 1981

Rodney Brooks, 'Unexpected Consequences of Self Driving Cars', 2017

Boer Deng, 'The Robot's Dilemma', *Nature,* 2015

Tobias Greitemeyer and Dirk O. Mügge, 'Video Games Do Affect Social Outcomes', *Personality and Social Psychology Bulletin,* 2014

Zoe Kleinman, '"Harmful" robot aims to spark AI debate', *BBC News,* 13 June 2016

'Principles of robotics', *EPSRC website*

Alexander Reben, 'Cool Hunting', via YouTube

Noel Sharkey, Aimee van Wynsberghe, Scott Robbins en Eleanor Hancock, 'Our Sexual Future with Robots', A Foundation for Responsible Robotics Consultation Report, 2017

Sherry Turkle, 'Connected, but alone?', *TED,* 2012

Sherry Turkle, *Alone Together – Why we expect more from technology and less from each other,* Basic Books, 2011

'A.I. is a crapshoot', *TV Tropes*

Alan Winfield, 'Responsible robotics (And what not to worry about)', lecture in De Balie, Amsterdam, 21 April 2015

Alan Winfield, 'Towards an Ethical Robot', 30 August 2014

Alan Winfield, 'The Gift', 29 December 2016

Chapter 12

Interviews
Ross D. King (18 July 2016), visit to robot scientist Eve, under the leadership of Chris Mellingwood in Manchester (26 July 2016), Robert Went (2 August 2017)

Literature and other media
AI100, '"Artificial Intelligence and Life in 2030." One Hundred Year Study on Artificial Intelligence: Report of the 2015-2016 Study Panel', Stanford University, September 2016
Nick Bostrom, 'Ethical Issues in Human Enhancement', *New Waves in Applied Ethics*, 2008
Ross D. King, 'Rise of the Robo Scientists', *Scientific American*, January 2011
Aimee Mullins, 'My 12 pairs of legs', *TED*, 2009
Jonathan Rossiter, 'Robotics, Smart Materials, and Their Future Impact for Humans', *BBVA Open Mind*, 2017
Michael Shermer, 'Artificial Intelligence Is Not a Threat – Yet', *Scientific American*, 1 March 2017
Alan Turing, 'Computing Machinery and Intelligence', Mind, 1950
Bruno van Wayenburg, *'Kijken: zo werken de molecuulmachines van Nobelprijswinnaar Ben Feringa'*, *NRC.nl*, 9 December 2016
Alan Winfield, 'Artificial intelligence will not turn into a Frankenstein's monster', *The Guardian*, 10 August 2014

Suggested reading

George Bekey, *Autonomous Robots – From Biological Inspiration to Implementation and Control*, MIT Press, 2005

Jack Copeland, *Artificial Intelligence – A Philosophical Introduction*, Wiley-Blackwell, 1993

Martin Ford, Rise of the Robots – *Technology and the Threat of a Jobless Future*, Basic Books, 2015

Maja Matarić, *The Robotics Primer*, MIT Press, 2007

Peter Menzel and Faith D'Aluisio, *Robo Sapiens – Evolution of a new species*, MIT Press 2001

Bennie Mols, Turings Tango – *Waarom de mens de computer de baas blijft*, Nieuw Amsterdam, 2012

Lisa Nocks, *The Robot – The Life Story of a Technology*, Greenwood Press, 2007

Lambèr Royakkers and Rinie van Est, *Just ordinary robots – Automation from love to war*, CRC Press, 2015

Bram Vermeer, Bennie Mols and Bas den Hond, *Robotics for Future Presidents*, TU Delft Robotics Institute, 2016

Bram Vermeer, Bennie Mols, Karin van den Boogaert, Jaco Boer and Bas den Hond, *Living with robots*, multimedia ebook, Oostenwind Publishing, 2016

Robert Went, Monique Kremer & André Knottnerus (red.), *Mastering the robot – The future of work in the second machine age*, WRR, 2015

Alan Winfield, *Robotics – A very short introduction*, Oxford University Press, 2012